SpringerBriefs in Environmental Science

SpringerBriefs in Environmental Science present concise summaries of cutting-edge research and practical applications across a wide spectrum of environmental fields, with fast turnaround time to publication. Featuring compact volumes of 50 to 125 pages, the series covers a range of content from professional to academic. Monographs of new material are considered for the SpringerBriefs in Environmental Science series.

Typical topics might include: a timely report of state-of-the-art analytical techniques, a bridge between new research results, as published in journal articles and a contextual literature review, a snapshot of a hot or emerging topic, an in-depth case study or technical example, a presentation of core concepts that students must understand in order to make independent contributions, best practices or protocols to be followed, a series of short case studies/debates highlighting a specific angle.

SpringerBriefs in Environmental Science allow authors to present their ideas and readers to absorb them with minimal time investment. Both solicited and unsolicited manuscripts are considered for publication.

More information about this series at http://www.springer.com/series/8868

Janet Hooke • Peter Sandercock

Combating Desertification and Land Degradation

Spatial Strategies Using Vegetation

With contributions by
G.G. Barberá, L. Borselli, C. Boix-Fayos,
L.H. Cammeraat, V. Castillo, S. De Baets, Janet Hooke,
J.P. Lesschen, M. Marchamalo, A. Meerkerk,
J.I. Querejeta, J.A. Navarro-Cano, J. Poesen,
Peter Sandercock, D. Torri, B. van Wesemael

Janet Hooke
Department of Geography and Planning
School of Environmental Sciences
University of Liverpool
Liverpool, UK

Peter Sandercock
Jacobs
Bendigo, VIC, Australia

Additional material to this book can be downloaded from http://extras.springer.com.

ISSN 2191-5547   ISSN 2191-5555   (electronic)
SpringerBriefs in Environmental Science
ISBN 978-3-319-44449-9   ISBN 978-3-319-44451-2   (eBook)
DOI 10.1007/978-3-319-44451-2

Library of Congress Control Number: 2016959455

Printed on acid-free paper

This Springer imprint is published by Springer Nature
The registered company is Springer International Publishing AG
The registered company address is: Gewerbestrasse 11, 6330 Cham, Switzerland

Acknowledgements

The RECONDES project was funded by the European Commission, Directorate-General of Research, Environment and Sustainable Development Programme, project no. GOCE-CT-2003-505361, FP6 programme 2004-7. The research was undertaken at the six institutions involved when authors were employed on the project: University of Portsmouth (UoP), UK (Janet Hooke, Peter Sandercock); Université Catholique De Louvain (UCL) (Bas Van Wesemael, Andre Meerkerk); Consiglio Nazionale Delle Ricerche – Istituto Di RicercaPer La Protezione Idrogeologica (CNR-IRPI) (Dino Torri, Lorenzo Borselli); Consejo Superior de Investigaciones Cientificas Centro de Edafologia y Biologia Aplicada del Segura (CSIC-CEBAS) (Victor Castillo, Gonzalo Barberá, J.A. Navarro-Cano); Universiteit van Amsterdam (UvA) (L.H. Cammeraat, Jan Peter Lesschen); Katholieke Universiteit Leuven (KUL) (Jean Poesen, Sarah De Baets).

The following sources of figures are gratefully acknowledged:

The project on which this publication is based was undertaken by personnel attached to six institutions. Each group had specific responsibilities according to land zone as well as collective contributions to the project and application of their specialisms. Authorship of sections, figures and tables relates to those responsibilities:

University of Portsmouth (UoP) (Hooke, Sandercock) – project leadership, channels, sub-catchments

CSIC-CEBAS, Spain (Barberá, Boix-Fayos, Castillo, Querejeta, Navarro-Cano) – reforested land

CNR-IRPI, Italy (Borselli, Torri) – catchment scale, modelling

UCL, Belgium (Meerkerk, van Wesemael) – rainfed cropland

UvA, Netherlands (Cammeraat, Lesschen) – semi-natural and abandoned land, field protocol, instrumentation

KUL, Belgium (De Baets, Poesen) – hillslopes, plant roots

Figures 2.3 and 5.2: Reprinted from Hooke, Sandercock, P (2012). Use of vegetation to combat desertification and land degradation: Recommendations and guidelines for spatial strategies in Mediterranean lands. Landscape and Urban Planning 107, 389–400, with permission from Elsevier.

Figures 2.8 and 2.9: Reprinted from Lesschen, J. P., Schoorl, J. M., & Cammeraat, L. H. (2009). Modelling runoff and erosion for a semi-arid catchment based on hydrological connectivity to integrate plot and hillslope scale influences. Geomorphology 109, 174–183, with permission from Elsevier.

Figure 3.2: Reprinted from: Ruiz-Navarro, A., Barberá, G. G., Navarro-Cano, J. A., & Castillo, V. M. (2009). Soil dynamics in Pinus halepensis reforestation: Effect of microenvironments and previous land use. Geoderma, 153(3–4), 353–361, with permission from Elsevier.

Figure 4.1: Reprinted from Meerkerk, A. L., van Wesemael, B., & Cammeraat, L. H. (2008). Water availability in almond orchards on marl soils in south east Spain: The role of evaporation and runoff. Journal of Arid Environments, 72, 2168–2178, with permission from Elsevier.

Figures 4.3 and 4.4: Reprinted from Verheijen, F., & Cammeraat, L. H. (2007). The association between three dominant shrub species and water repellent soils along a range of soil moisture contents in semi-arid Spain. Hydrological Processes, 21, 2310–2316, with permission from John Wiley and Sons. Copyright © 2007 John Wiley & Sons, Ltd.

Contents

Contributors

Gonzalo Barberá Centro de Edafologia y Biologia Aplicada del Segura (CEBAS), Department of Soil and Water Conservation, Campus Universitario de Espinardo, Murcia, Spain

Carolina Boix-Fayos Centro de Edafologia y Biologia Aplicada del Segura (CEBAS), Department of Soil and Water Conservation, Campus Universitario de Espinardo, Murcia, Spain

Lorenzo Borselli Institute of Geology/Faculty of Engineering, Universidad Autonoma de San Luis Potosì (UASLP), San Luis Potosí, Mexico

L.H. Cammeraat Instituut voor Biodiversiteit en Ecosysteem Dynamica (IBED) Earth Surface Science, Universiteit van Amsterdam, Amsterdam, The Netherlands

Victor Castillo Centro de Edafologia y Biologia Aplicada del Segura (CEBAS), Department of Soil and Water Conservation, Campus Universitario de Espinardo, Murcia, Spain

Sarah De Baets College of Life and Environmental Sciences, Department of Geography, University of Exeter, Exeter, UK

Janet Hooke Department of Geography and Planning, School of Environmental Sciences, University of Liverpool, Liverpool, UK

Jan Peter Lesschen Alterra, Wageningen University and Research Centre, Wageningen, The Netherlands

Miguel Marchamalo Departamento de Ingenieria y Morfologia del Terreno, Universidad Politecnica de Madrid, Madrid, Spain

André Meerkerk Earth and Life Institute, Université catholique de Louvain (UCL), Louvain-la-Neuve, Belgium

J.A. Navarro-Cano Centro de Investigaciones sobre Desertificacion (CSIC-UV-GV), Valencia, Spain

Jean Poesen Division of Geography and Tourism, Department of Earth and Environmental Sciences, KU Leuven, Heverlee, Belgium

J.I. Querejeta Centro de Edafologia y Biologia Aplicada del Segura (CEBAS), Department of Soil and Water Conservation, Campus Universitario de Espinardo, Murcia, Spain

Peter Sandercock Jacobs, Bendigo, VIC, Australia

Dino Torri Consiglio Nazionale Della Ricerche – Istituto di Ricerca per la Protezione Idrogeologica (CNR-IRPI), Perugia, Italy

Bas van Wesemael Earth and Life Institute, Université catholique de Louvain (UCL), Louvain-la-Neuve, Belgium

Chapter 1
Introduction

Janet Hooke, Gonzalo Barberá, L.H. Cammeraat, Victor Castillo, Jean Poesen, Dino Torri, and Bas van Wesemael

Abstract This book explains the methods and results of a major research project, RECONDES, that was undertaken to develop strategies of effective use of vegetation to combat desertification and land degradation by water. The research approach combined understanding of the processes of erosion and land degradation with identification of suitable and effective plants and types of vegetation that could be used to decrease the intensity of soil erosion. The project uses the relatively new concept of physical connectivity of water and sediment in the landscape. The premise of the approach is that sediment connectivity can be reduced through the development of vegetation in the flow pathways, and that this approach is more sustainable than use of physical structures. It required research into the locations and characteristics of these pathways and into properties of suitable plants and species at a range of scales and land units. These components are combined to produce a spatial

J. Hooke (✉)
Department of Geography and Planning, School of Environmental Sciences,
University of Liverpool, Roxby Building, L69 7ZT Liverpool, UK
e-mail: janet.hooke@liverpool.ac.uk

G. Barberá • V. Castillo
Centro de Edafologia y Biologia Aplicada del Segura (CEBAS),
Department of Soil and Water Conservation, Campus Universitario
de Espinardo, PO Box 164, 30100 Murcia, Spain

L.H. Cammeraat
Instituut voor Biodiversiteit en Ecosysteem Dynamica (IBED)
Earth Surface Science, Universiteit van Amsterdam, Science Park 904,
1098 XH Amsterdam, The Netherlands

J. Poesen
Division of Geography and Tourism, Department of Earth and Environmental Sciences,
KU Leuven, Celestijnenlaan 200E, 3001 Heverlee, Belgium

D. Torri
Consiglio Nazionale Della Ricerche – Istituto di Ricerca per la Protezione Idrogeologica
(CNR-IRPI), Via Madonna Alta 126, Perugia, Italy

B. van Wesemael
Earth and Life Institute, Université catholique de Louvain (UCL),
Place Louis Pasteur 3, 1348 Louvain-la-Neuve, Belgium

© Springer International Publishing Switzerland 2017   1
J. Hooke, P. Sandercock, *Combating Desertification and Land Degradation*,
SpringerBriefs in Environmental Science, DOI 10.1007/978-3-319-44451-2_1

strategy of use of suitable plants at the most strategic points in the landscape, designed for restoration or mitigation of land degradation. Additional benefits of use of vegetation as a strategy of sustainable management are outlined. The methods and restoration strategy were developed in relation to the dryland environments of the Mediterranean region of southern Europe, involving field measurements, monitoring and modelling in the study area in Southeast Spain, the driest and most vulnerable region in Europe to desertification.

Keywords Research methods • Sustainable land management • Landscape approach • Vegetation restoration • Soil erosion control • Catchment monitoring • Landscape connectivity

1.1   Context and Problem

Desertification and land degradation are major problems worldwide, with semi-arid and sub-humid lands particularly vulnerable. The extent of the areas affected and the severity of the problem are predicted to increase in the future under most land use and climate change scenarios (Lesschen et al. 2007; Safriel and Adeel 2005; Zollo et al. 2015). It is therefore an urgent and serious challenge to design and implement effective strategies and methods to combat desertification and land degradation.

Desertification is taken to mean the process of becoming desert-like (it therefore does not include the major arid land deserts that are 'naturally' and have long been deserts). In many regions the main processes of land degradation involve erosion by water. Elsewhere, especially in areas adjacent to sandy deserts such as in China, the major soil erosion processes are by wind, and deposition takes place as sand dunes, which may move and encroach on productive areas. In the processes produced by flowing water, major amounts of topsoil removal occur, leading to decreased soil quality, nutrient loss and reduced infiltration. These in turn produce a positive feedback by increasing runoff and hence further accelerating soil erosion. The soil material is transported down through the catchment, where it silts up reservoirs, or causes muddy flows during flooding. Overall, severe soil erosion and gullying of land result in loss of agricultural productivity, thus increasing poverty, migration and other social problems (Millenium Ecosystem Assessment 2005). Land degradation through these processes produces many negative 'on-site' and 'off-site' effects, especially during intense seasonal rainfall events.

It has long been recognised that the main way to combat desertification/land degradation is by increasing or at least maintaining vegetation cover (Millenium Ecosystem Assessment 2005). Conventional solutions for mitigating desertification include reforestation of large areas and the construction of check dams, both of which have long been used in some regions but which are costly. They also have detrimental effects and may not be that effective (Boix-Fayos et al. 2007; Boix-Fayos et al. 2008; Brown 1944). Soil conservation practices have also long been advocated and a wealth of literature exists on soil conservation techniques such as terracing, contour ploughing, use of grass strips, soil treatments and tillage strategies (Morgan 2005).

The purpose of this publication is to report a major research project that was undertaken to develop strategies of more effective use of vegetation to combat desertification and land degradation by water (RECONDES, http://www.port.ac.uk/research/recondes/). It demonstrates the approaches, methods and results in order to exemplify their potential wider application in other environments and in practical schemes. The approach was developed in relation to the Mediterranean environment but it has much wider potential applicability to dryland and vulnerable areas. Major problems with present approaches to mitigation or restoration in drylands are that either they are a blanket approach, as in afforestation covering whole hillsides or catchments, or piecemeal as in use of particular structures, or only work in the short-term, as with check dams. Some traditional and ancient practices are effective, for example agricultural terracing, but some of these are falling into disuse. The need was therefore to develop strategies that combine understanding of the processes of erosion and land degradation with identification of plants and types of vegetation that could be used to decrease the intensity of these processes. A major premise of the project is that much of the erosion and soil removal and its subsequent deposition occurs in limited pathways within the landscape and therefore these should be targeted. If vegetation is placed in these pathways then it can reduce erosion and transmission of effects but does not occupy all the land, so agricultural land can remain productive. The project uses the relatively new concept of understanding physical connectivity of water and sediment in the landscape. It required research into the locations and characteristics of these pathways and into properties of suitable plants and species. Another major premise of the project was that only indigenous species should be used in these strategies, because of problems caused by exotic species introduced to areas (see D'Antonio and Meyerson 2002, and Rodríguez 2006 for a review of the pros and cons and controversies about the use of exotic species in restoration).

In semi-arid areas, including the driest parts of Mediterranean Europe, erosion has still been detected in reforested areas as a result of a combination of poor tree growth, low ground cover and the terrace structures created for the forestry (Castillo et al. 2001). Check dams tend to provide only short-term solutions because they fill up very quickly in such environments (Cammeraat 2004; Lesschen et al. 2008a), and they are often breached, even by quite moderate storms (Borselli et al. 2006; Castillo et al. 2007; Hooke and Mant 2001, 2002; Poesen and Hooke 1997). Soil erosion on agricultural lands is the major problem (Cerdà et al. 2009) but modern land practice may also increase erosion risk by opening up long slopes and leaving land bare for periods (Borselli et al. 2006). These problems are likely to be exacerbated in many dryland and sub-humid areas in the future because of global climate change (IPCC 2007). A whole-landscape approach has been promoted in Belgium (AMINAL 2002; Evrard et al. 2007) but no attempt had been made, to our knowledge in 2004, to develop this approach in a Mediterranean environment, with its particular plants and cultural landscape features, notably agricultural terracing. Some knowledge on aspects of Mediterranean vegetation and plants existed at the time of the initiation of this project (Hooke 2007b) but detailed work on plant properties, growth requirements and response to processes was limited (Sandercock et al. 2007). Some work was ongoing into the use of plants for revegetating road cuttings in the region (Bochet et al. 2007; Tormo et al. 2007), since these are also major problem zones (Cerdà 2007), and for slope stabilisation (e.g. Mattia et al. 2005). Research on restoration of gullied and degraded areas includes experimental plantings in the Mediterranean Alps (Burylo et al. 2007) and much work is ongoing into other techniques to be applied in the Mediterranean region (e.g. García-Orenes et al. 2009; Giménez Morera et al. 2010). The present state of knowledge on soil erosion in the Mediterranean region is reviewed in Cerdà et al. (2010). Research on these various aspects is accelerating.

1.2 Processes and Connectivity Concept

Most of the soil erosion and land degradation in the Mediterranean region occurs by the movement of surface material downslope via water flow transport in rills, gullies and soil tunnels and micropipes. Erosion is greatest where flow velocities or shear stresses are highest, combined with low resistance of the top soil or surface, i.e. on steep gradient areas, and where runoff generation is greatest, notably bare surfaces. These are erosion hotspots. Erosion is much reduced where the vegetation cover is >30 % (Thornes and Brandt 1993). Once gullies or piping develop then these can be propagated rapidly up and downslope.

Conventional analyses relate erosion rates or sediment yield directly to catchment characteristics through statistical analysis or are derived from plot scale studies. More recently, the concept of connectivity has been applied to understand sediment flux in catchments, as well as to hydrological analysis (Bracken and Croke 2007; Reaney et al. 2007). Although this approach was developed in Australia (Brierley and Fryirs 1998; Brierley and Stankoviansky 2002; Fryirs and Brierley 2000; Fryirs et al. 2007b), it has been tested in Spain and Luxembourg (Cammeraat 2002; Imeson and Prinsen 2004), and in relation to coarse sediment connectivity in stream channels (Hooke 2003). Connectivity is defined in this context as "the physical linkage of water or sediment flux within the landscape and the potential for a particle to move through the system" (Hooke 2003). The importance of connectivity at local and hillslope scale, particularly the effects of vegetation patchiness in Mediterranean and semi-arid landscapes, was long ago recognised (e.g. Cammeraat and Imeson 1999; Cerdà 1997; Puigdefabregas et al. 1999) and such ideas are being used in practical management in Australia, following the important work of Ludwig and Tongway (2000) Ludwig et al. (2002) and Tongway et al. (2003). The catchment scale research on connectivity (e.g. Fryirs et al. 2007a; Quiñonero-Rubio et al. 2013) has demonstrated that major areas of storage can occur within the landscape, reducing the transmission of the sediment load downstream. It focuses on the major patterns of sediment removal some of which are missed by plot scale studies. It identifies pathways and 'hotspots' of erosion. The concept of connectivity was a major tenet of the RECONDES project and resulted in various methodological developments and methods of mapping, constituting a rich and varied series of contributions to the literature on the topic (Borselli et al. 2008; Lesschen et al. 2008a; Lesschen et al. 2008b; Lesschen et al. 2009; Marchamalo et al. 2016; Meerkerk et al. 2009; Sandercock and Hooke 2011; Vigiak et al. 2012).

1.3 Benefits of Use of Vegetation

Vegetation has several effects which are advantageous in reducing surface runoff of water (and therefore also making it more available to plants by infiltration into the soil) and reducing movement of soil and nutrients (Table 1.1). The advantages of using vegetation in a spatially strategic approach mean that it should be more sustainable and efficient than conventional approaches. With use of suitable plants then they should be self-sustaining. The major advantage of the spatially strategic approach, that only places vegetation in particular locations such as field boundaries, is that it is does not take up large areas of productive land and the budget for restoration may be much reduced to obtain an optimal or near optimal erosion reduction. This approach also increases spatial connectivity for wildlife, contributing to biodiversity conservation. Furthermore, in the longer-term this favours the formation of a soil biota, mainly fungi and michorrizae, whose interaction with plants produces soil organic matter (Rillig et al. 2015). Soil amelioration also improves ecosystem resilience, which is of outstanding importance due to ongoing

Table 1.1 Beneficial effects of vegetation in soil erosion and land degradation reduction

Feature	Effect(s)
Vegetation canopy	Protects the soil surface from the impact of rain drops, and direct solar radiation; both benefit biological activity and soil and aggregate formation processes
Vegetation structure (Stems & leaves)	Increases rugosity (roughness) and so lowers flow velocities and sediment transport; this will increase sedimentation
Root length density	Increases soil cohesion and hence resistance to erosion
Root-soil interface	Increased root length density increases porosity and macro-porosity thus promoting infiltration so increasing water availability to plants locally and decreasing flow downstream, ultimately helping reduce flood risk; interaction with soil biota favours soil aggregate stability, improving soil pore system stability and decreasing soil crust formation
Soil quality	Increases organic matter, texture, nutrient, water retention and microbiological activity
Carbon sequestration	More generally, sequesters carbon so mitigating greenhouse gas emissions and thus global warming impacts

climate evolution. Furthermore, favourable soil physical properties are maintained or improve under developing or permanent vegetation, enhancing soil porosity and infiltration. The approach developed and advocated in this project and publication is therefore a win-win situation.

1.4   Approach

The premise of the approach developed here is that sediment connectivity can be reduced through the development of vegetation in the flow pathways, and that this approach is more sustainable than use of physical structures (Sandercock and Hooke 2006). The approach requires the combination of understanding of the physical processes, particularly the connectivity pathways, and a knowledge of the suitability and effectiveness of plants to reduce erosion. These components are combined to produce a spatial strategy of use of suitable plants at the most strategic points in the landscape, designed for restoration or mitigation of land degradation in the dryland environments of the Mediterranean region of southern Europe. The steps in application of this approach are shown in Fig. 1.1. This strategy is designed (a) for application in areas which are prone to soil erosion by water, where vegetation is sparse, and primarily with agricultural land use, and (b) to 'work with nature' using indigenous and local plant species. The Mediterranean areas have a long tradition of agriculture where agricultural terraces have been built on steep slopes that are highly vulnerable to land degradation, especially where on soft bedrock such as marl. The use of local native species means that they are adapted to the local environmental conditions, and avoids problems of introducing exotic species. The general characteristics of suitable plants to be effective and that should guide selection of species are outlined in Table 1.2.

The development of this methodology and the provision of necessary information about suitable types of plants required fundamental research into three components (Fig. 1.2):

1. identification of the prime locations for targeting revegetation, i.e. locate the erosion hotspots, flow pathways, and measure connectivity in the landscape at various scales;
2. identification of what plant species grow within the catchment and specifically in the erosion hotspots, then identification of the conditions required for their growth;
3. assessment and selection of the most effective plant species for establishing in the landscape to control erosion and increase sedimentation in specific locations.

The RECONDES project was developed to address these aims and undertake the research. It was funded by the European Commission in its Framework 6 programme (project no. GOCE-CT-2003-505361) as part of its concern with future effects of global and climate change and likely increased desertification threatening

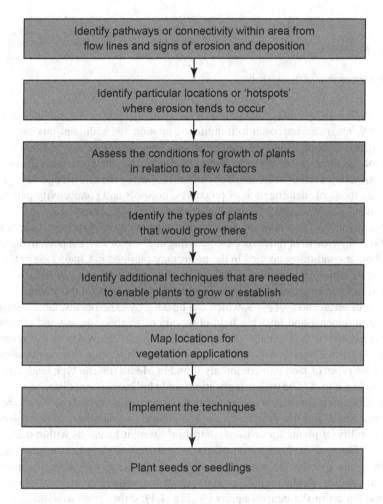

Fig. 1.1 Flow diagram of method of application of spatial vegetation strategy (After Hooke and Sandercock 2012)

Table 1.2 Desirable plant characteristics for erosion control

Properties
Native species
Germinates and/or propagates easily
Rapid growth rate
Perennial or persistent
Ability to grow in a range of substrates
Drought tolerant
Produces a dense root system
Has a high threshold to withstand forces of water flow
Ability to trap water and sediments

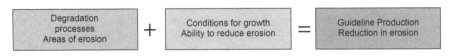

Fig. 1.2 Components of the research

Europe. A major objective of the project was to produce practical guidelines of how a spatially integrated approach to minimising erosion and sediment flux and mitigating desertification could be implemented. This was done by combining the results of the three research components. Methods have therefore been developed for investigating each of these components and for using the knowledge acquired.

The methods of identification of processes, hotspots and connectivity pathways and the results of measurements and analysis in this project are explained in Chap. 2. The identification of potential species growing in the area and their types of location, conditions for growth of those species including analysis of soil, edaphic, hydrological and other conditions, applied in the project are outlined in Chap. 3. Selection of the most suitable plants involves assessment of their effectiveness in reducing erosion and in increasing sedimentation. This may involve tests on the strength of the plants, both aerial and root parts, which can be done by experiments, laboratory and field measurements and involves field observations of the response and effects of plants in relation to flow events. The methods and results derived from this research are described in Chap. 4. A synthesis of the results illustrating their application in the various types of land unit commonly found in Mediterranean type landscapes is provided in Chap. 5. The wider implications and applicability are discussed.

The fundamental research on each of these aspects was carried out for an area in Southeast Spain and has produced new knowledge on conditions of plant growth and suitability of plants for erosion control in different locations within that environment. The specific findings on plants can be applied elsewhere in that region, with the same species, but wider application of this strategy in other regions with different species would require assessment of the suitability of local plants for implementation in the general approach (Fig. 1.1), either from available existing knowledge or from similar research to that outlined here. The purpose of this Springer Brief is therefore to explain the research methods adopted and to illustrate the research results and then to demonstrate the application of this approach and the production of the practical guidelines. Such research methods and this approach to land management could be implemented in other areas of the world vulnerable to desertification and land degradation.

1.5 Research Design and Study Area

Most of the research in RECONDES has been carried out in Southeast Spain, this being the driest and most vulnerable region in Europe to desertification. Cárcavo catchment in the northeast of the Region of Murcia (Fig. 1.3) was selected as the

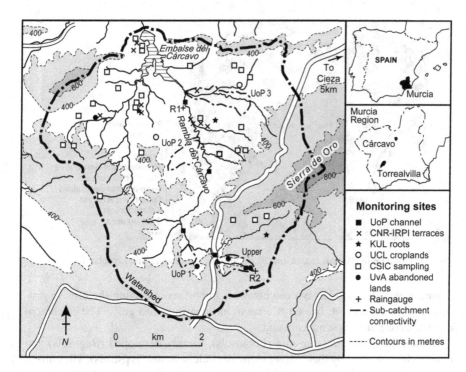

Fig. 1.3 Location of and characteristics of Cárcavo catchment and position of study sites and instrumentation

main catchment for all field work so as to maximise synergies and because it had the properties and conditions required (e.g. mixed land uses, marl lithology, extensive degraded land, management through forestation and check dams). Some work has also been completed in other catchments in Southeast Spain that had been previously studied in earlier projects and where long-term monitoring is still being carried out, particularly the Torrealvilla catchment in the Guadalentín basin in the southwest of the Region of Murcia.

1.5.1   General Catchment Characteristics

The Cárcavo basin is a small catchment of 30 km² and an altitude range between 220 and 850 m. This region of Spain is very dry with an average annual rainfall of 300 mm and a potential evapotranspiration of 900 mm. The geology of the area consists of steep Jurassic limestone and dolomite mountains with calcareous piedmonts, surrounding deposits of Cretaceous and Miocene marls, and Keuper gypsum deposits. The outlet of the basin drains directly into the Segura River, the major river system of southeastern Spain. The large difference of base level between the

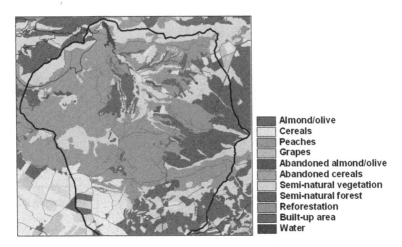

Almond/olive
Cereals
Peaches
Grapes
Abandoned almond/olive
Abandoned cereals
Semi-natural vegetation
Semi-natural forest
Reforestation
Built-up area
Water

Fig. 1.4 Land use in Cárcavo catchment, 2004 (Authors: Cammeraat and Lesschen)

Segura River and the Cárcavo catchment is important in driving the erosive processes in the catchment. Indeed 'Cárcavo' in Spanish means gully. The catchment has a reservoir at its downstream end.

Most soils in the area are thin (Leptosols), weakly developed (Regosols) and mainly characterized by their parent material (Calcisols and Gypsisols). The current land use in the study area consists of cereals, grapes, olives, almonds, abandoned land, reforested land and semi-natural vegetation. In the 1970s large parts of degraded land were reforested with Aleppo pine (*Pinus halepensis* Mill.) within the framework of reforestation and soil conservation programmes. Some almond and olive fields in the central part are under irrigation, while low-yielding cereals are found on marls without irrigation. During the last 50 years, parts of the rainfed agriculture have been abandoned and are under different stages of secondary succession. The steeper and higher areas are under semi-natural vegetation and, on north slopes or gentle pediments, under forest.

Land use was mapped using digitisation of air photos and by field survey. Land use change was analysed from aerial photographs of 1985 and 1997. For 2004, the land use map of 1997 was updated during fieldwork in September 2004. Eleven different land use classes were distinguished of which six are related to agriculture. The created land use map for 2004 is shown in Fig. 1.4.

As can be seen from Fig. 1.4, a major part, especially the southeast part of the study area, is now almost completely under almond/olive, having replaced cereals in the period 1984–2004, with the area of abandoned cereals and almond/olive almost doubling during that time. Almost no land use changes occurred in the semi-natural and reforested areas during the last 30 years or more, since the reforestation programs are mainly from the 1970s.

Detailed sites for measurements and monitoring were set up within the Cárcavo catchment; the location of these sites is shown for different components of the study in Fig. 1.3. After the selection of field sites for specific purposes, equipment for monitoring runoff, sediment delivery and soil moisture was installed. At the main field site in the upper part of the catchment (Site 2, R2 on Fig.1.3) a tipping bucket, six FDR soil moisture sensors, three temperature sensors, two pressure transducers and several gypsum sensors, all connected to a data logger, were installed. Furthermore, 12 runoff indicators and two runoff gutters were placed at this site. One stand-alone pressure transducer was installed in the upper part of the Cárcavo channel, another one in a major tributary channel and a third in the lower part of the Cárcavo channel (Fig. 1.3), connected to a data logger and a tipping bucket rain gauge. Additionally, field sites in the Guadalentín basin in the south of Murcia province were being maintained, in the Torrealvilla basin and the Alqueria catchment where already a long term data record existed with regard to rainfall-runoff relationships at various scales (Cammeraat 2004; Cammeraat et al. 2009; Hooke 2007a). As rainfall events normally are very localised, more study areas where connectivity and runoff patterns were being monitored was considered useful.

1.5.2 Monitoring Programme and Hydrological Conditions

This research, involving field measurements and experiments, was carried out in the period 2004–2007. Below are details of the rainfall conditions in that period because this affected the number of events and the results it was possible to obtain.

In the Cárcavo basin two sites were maintained where rainfall was measured at 1-min interval. One site was located in the lower part of the catchment next to the main Cárcavo channel and the other is located in the southeast, in the upper part of the catchment at the upper abandoned lands site (Fig. 1.3). Rainfall was measured for more than 2 years and some statistics are given in Table 1.3. In general, the rainfall at the upper site is about 7 % higher than at the channel site, probably because of the higher altitude and rain shadow effect of the Sierra del Oro. The year 2005 was dry in Cárcavo with about 235 mm (20 % under the average) and very dry in the Guadalentín site with only 168 mm. In 2006 rainfall was about 50 % above the average in Cárcavo the deviation attributable to 160 mm of rain falling in the first days of November, as recorded at Site 2. Although the distance between the two measuring sites is only 4 km, the amount of rainfall recorded at each site varied considerably (Table 1.3).

For these events, in many cases there was a difference of at least 5 mm between the upper and lower parts, mainly related to autumn rainfall events, which are often torrential events with a high spatial variability (González-Hidalgo et al. 2001). For modelling purposes the spatial distribution of the rainfall is therefore very important, since no uniform rainfall can be expected for the whole catchment. This was also observed for the November 2006 event, where most erosion was observed in the southern (upper) part of the catchment.

During the first days of November 2006 in total about 160 mm of rainfall was measured at the rain gauge in the upper part of the catchment (Site 2). Most of this rainfall was of low intensity, only the last rainfall on 8 November was of high inten-

Table 1.3 Rainfall characteristics and statistics for both measuring sites in Cárcavo

	Channel site	Upper site
Altitude of rain gauge (m)	276	421
Sum rainfall 2005 (mm)	230	243
Sum rainfall 2006 (mm)	312[a]	446
Sum rainfall 2007 (till 19 March) (mm)	106	117
Sum of rainfall without missing data (mm)	482	514
Number of events of >1 mm	63	72
Number of events of >10 mm	13	15
Number of events of >20 mm	7	6

[a]Due to malfunctioning of the rain gauge the November 2006 event is not recorded at the channel site, and therefore total rainfall is underestimated

sity with 37 mm within 3 h and a maximum intensity of 50 mm/h within a 10 min interval. The recurrence time of the event of 8 November was about 2 years, based on a 36-year data series of daily rainfall at the Almadenes weather station, located just north of the catchment near Cieza. However, when the rainfall of the previous days was taken into account the recurrence time for the 10-day rainfall sum was 23 years. Since the soil was already saturated after the rainfall of the previous days this event caused significant erosion, which was observed by completely filled sediment boxes, a reaction of all runoff indicators, rills on agricultural fields, failure of some terraces, and a flow in the lower part of the channel of approximately 60 cm height.

If the rainfall for the study period 2004–2007 is compared with the longer-term record the project period is seen as particularly dry (Hooke and Mant 2015). Certainly the total rainfall in the whole region was below average for the whole study period. This did restrict the number of rainfall events it was possible to measure and particularly the number of intense rainfall events, the November 2006 event being the exception. The infrequent occurrence of rainfall events in semi-arid areas and their localised spatial extent is a problem for all such monitoring and research programmes. Ideally, extended periods or monitoring of several catchments is needed but this is often not feasible in research programmes.

1.6   Conclusion

This chapter has set out the major premises of the approach developed to use vegetation to mitigate desertification and reduce soil erosion and land degradation by reduction of connectivity in the landscape. It has explained the overall research design used and the mode of application in the study area in Southeast Spain. The following three chapters explain the methods and results for each of the major components of the research. These are synthesised and developed into practical recommendations in Chap. 5.

Chapter 2
Mechanisms of Degradation and Identification of Connectivity and Erosion Hotspots

Janet Hooke, Peter Sandercock, L.H. Cammeraat, Jan Peter Lesschen,
Lorenzo Borselli, Dino Torri, André Meerkerk, Bas van Wesemael,
Miguel Marchamalo, Gonzalo Barbera, Carolina Boix-Fayos, Victor Castillo,
and J.A. Navarro-Cano

Abstract The context of processes and characteristics of soil erosion and land degradation in Mediterranean lands is outlined. The concept of connectivity is explained. The remainder of the chapter demonstrates development of methods of mapping, analysis and modelling of connectivity to produce a spatial framework for development of strategies of use of vegetation to reduce soil erosion and land degradation. The approach is applied in a range of typical land use types and at a hierarchy of scale from land unit to catchment. Patterns of connectivity and factors influencing the location and intensity of processes are identified, including the influence of topography, structures such as agricultural terraces and check dams, and past land uses. Functioning of connectivity pathways in various rainstorms is assessed. Modes of terrace construction and extent of maintenance, as well as presence of tracks and steep gradients are found to be of importance. A method of connectivity modelling that incorporates effects of structure and vegetation was developed and has been widely applied subsequently.

J. Hooke (✉)
Department of Geography and Planning, School of Environmental Sciences,
University of Liverpool, Roxby Building, L69 7ZT Liverpool, UK
e-mail: janet.hooke@liverpool.ac.uk

P. Sandercock
Jacobs, 80A Mitchell St, PO Box 952, Bendigo, VIC, Australia
e-mail: Peter.Sandercock@jacobs.com

L.H. Cammeraat
Instituut voor Biodiversiteit en Ecosysteem Dynamica (IBED) Earth Surface Science,
Universiteit van Amsterdam, Science Park 904, 1098 XH Amsterdam, The Netherlands

J.P. Lesschen
Alterra, Wageningen University and Research Centre, Droevendaalsesteeg 3 (GAIA),
PO Box 47, 6700AA Wageningen, The Netherlands

L. Borselli
Institute of Geology/Faculty of Engineering, Universidad Autonoma de San Luis Potosì
(UASLP), Av. Dr. Manuel Nava 5, 78240 San Luis Potosí, S.L.P., Mexico

© Springer International Publishing Switzerland 2017   13
J. Hooke, P. Sandercock, *Combating Desertification and Land Degradation*,
SpringerBriefs in Environmental Science, DOI 10.1007/978-3-319-44451-2_2

Keywords Soil erosion processes • Runoff connectivity • Sediment connectivity • Erosion hotspots • Agricultural terraces • Connectivity mapping • Connectivity modelling

2.1   Soil Erosion and Degradation in Desertified Mediterranean Lands

The Mediterranean environment is affected by desertification. Desertification means land degradation in arid, semi-arid and dry sub-humid areas resulting from various factors, including climatic variations and human activities (UNCCD 1994). Of all the processes leading to desertification, soil erosion is considered to be the most important degradation process in Mediterranean Europe (Poesen et al. 2003). Erosion and degradation occur across a range of scales and land units in the Mediterranean landscape. Within each of these scales and land units it is possible to identify particular processes of erosion and land use practices which have directly or indirectly contributed to desertification in the region.

Erosion is particularly important in agro-ecosystems of marginal hilly areas characterized by high water deficit and high inter-annual variability in rainfall. The change in land use from a semi-natural vegetation or traditional terraced orchards to intensified plantations with an extremely low crop cover, leaving most of the soil exposed, has large impacts on hydrology and land degradation (Beaufoy 2002; De Graaff and Eppink 1999), with both on-site effects such as gullying and off-site effects such as flooding and reservoir sedimentation.

D. Torri
Consiglio Nazionale Della Ricerche – Istituto di Ricerca per la Protezione Idrogeologica (CNR-IRPI), Via Madonna Alta 126, Perugia, Italy

A. Meerkerk • B. van Wesemael
Earth and Life Institute, Université catholique de Louvain (UCL),
Place Louis Pasteur 3, 1348 Louvain-la-Neuve, Belgium

M. Marchamalo
Departamento de Ingenieria y Morfologia del Terreno, Universidad Politecnica de Madrid, Madrid, Spain

G. Barbera • C. Boix-Fayos • V. Castillo
Centro de Edafologia y Biologia Aplicada del Segura (CEBAS),
Department of Soil and Water Conservation, Campus Universitario
de Espinardo, PO Box 164, 30100 Murcia, Spain

J.A. Navarro-Cano
Centro de Investigaciones sobre Desertificacion (CSIC-UV-GV),
Carretera Moncada – Náquera, Km. 4,5, 46113 Moncada, Valencia, Spain

In semi-natural and abandoned lands, disturbances to the spatial heterogeneity of vegetation cover result in concentrated flow (Cammeraat and Imeson 1999; Tongway and Ludwig 1996) leading to high degradation rates and soil quality loss. Many abandoned lands are also terraced by earth dams and with the halt of cultivation, the maintenance of terraces is also stopped. This leads to the deterioration of the terraces, an increase in the length of slopes over which runoff occurs and accelerated rates of erosion. Soil retained above these terraces may be eroded and released under extreme runoff events (Cammeraat 2002).

Reforestation using trees has been widely applied throughout the Mediterranean; however, due to the harshness of semi-arid environments their success has been severely limited (Zhang et al. 2002). Extensive terracing of lands was undertaken across the Mediterranean in the 1960s and 1970s using heavy machinery, a practice considered to improve water yield to the plants and accelerate development of vegetation and ecosystem restoration. However, in many cases, land degradation was triggered by these aggressive techniques, increasing soil erosion (Chaparro and Esteve 1995; Williams et al. 1995) and reducing soil quality (Querejeta 1998). Castillo et al. (2001) indicate that the sidebanks produced may also be a source of sediment and gully initiation because of the long-term devegetation and steep gradient.

Gully erosion is a particular problem, responsible for significant on-site soil losses and off-site consequences (Poesen and Valentin 2003). In Mediterranean areas, the evidence from several studies is that gully erosion may be responsible for up to 80% of total soil losses due to water erosion, whereas this process in many cases only operates on less than 5% of the land area (Poesen et al. 2003). Once developed, gullies increase the connectivity of flow and sediment transfers from uplands to lowland areas through the drainage system. Reducing gully erosion, will lead to less sediment export, less reservoir sedimentation, lower flood risk and allow more water in uplands to infiltrate. Similarly, deterioration of soil resources and ecosystems is also apparent in valley floors, mainly as a result of erosion during high flows. Efforts that are made to mitigate runoff and concentrated flow leading to gully erosion would also be effective in reducing the potential for erosion and degradation of valley floors.

2.2 Processes

Most of the soil erosion and land degradation in the Mediterranean region occurs by the movement of surface material downslope via water flow transport in rills, gullies and soil pipes. Rainsplash can have an effect on bare soil and lead to crusting, which further decreases infiltration rates and increases runoff. Some erosion may take place in overland flow in unrilled zones at the top of slopes but quickly water flow tends to be concentrated into pathways which then incise at some distance downslope producing rills. Erosion is greatest where flow velocities or shear stresses are highest, combined with low resistance of the soil or surface, i.e. on steep gradient areas, and where runoff generation is greatest, notably bare surfaces. These are erosion hotspots. Erosion is much reduced where the vegetation cover is >30% (Thornes and Brandt 1993). Amount of runoff increases downslope and thus also length of slope increases liability to erosion. Once gullies or piping develop then these can be propagated rapidly up and downslope. For all these reasons increase of vegetation and particularly emplacing vegetation in the pathways of flow can reduce soil erosion and sediment flux. Infiltration rate of the surface is a major control and once soil erosion begins then soil removal tends to decrease infiltration rate, which increases runoff which in turn increases erosion and so decreases infiltration rate still further. This positive feedback relation can proceed until all soil is removed and more resistant bedrock is reached. Agricultural productivity of such areas becomes rapidly reduced. Thus soil erosion tends to accelerate once initiated.

Ruiz-Navarro et al. (2012) studied the relationship between landscape attributes and soil quality in Cárcavo catchment. Soil quality is clearly structured in a gradient of fertility/quality where soil carbon, nutrient availability, cation exchange capacity, relatively low pH and water holding capacity are strongly associated. Higher quality soils are positively associated to landscape configurations of flow convergence and low solar radiation (low hydric stress) and dense vegetation and negatively associated to areas of intense gully formation on highly erodible lithologies and scarce vegetation (active erosion processes). They also found that the positive correlation to dense vegetation and water favourable conditions is higher at finer resolution (5–10 m) while the negative association to erosion-prone configurations is higher at coarser resolution (20–40 m). This is interesting as the topographic structures favouring flow convergence simultaneously promote vegetation development and erosion processes, and, depending on the connectivity of flows, the balance may go one way (vegetation and higher soil quality) or the other (erosion and poor soil quality), this being consistent with better association of positive scenarios (vegetation predominance) at finer scales (flows have not enough energy to connect) and negative scenarios (erosion dominance) at coarser scales (flows have higher energy and overcome vegetation resistance to connection). The system would act as a tipping point threshold with alternative states of vegetation positive feedback or erosion positive feedback.

2.3 Connectivity Concept and Methods

When the RECONDES project was designed very little work applying the connectivity concept to the Mediterranean environment and to practical uses had been undertaken. Methods of mapping, identification, and quantification were still poorly developed, with a few exceptions e.g. van Dijk et al. (2005) and the work of Cammeraat and Imeson (1999), Cammeraat (2002), Cammeraat (2004) on hillslopes and Hooke (2003) in channels. Since then much development has taken place, particularly in the conceptual ideas, including ideas of structural and functional connectivity (Lexartza-Artza and Wainwright 2009; Okin et al. 2009; Wainwright et al. 2011), and in relation to applications (e.g. Brierley et al. 2006;

Croke et al. 2005; Evrard et al. 2007). Much discussion on runoff and hydrological connectivity has taken place (Bracken et al. 2013). Various methodological developments and methods of mapping arose from the RECONDES project which produced a very rich and varied series of contributions (Borselli et al. 2008; Lesschen et al. 2008a, b, 2009; Marchamalo et al. 2016; Meerkerk et al. 2009; Sandercock and Hooke 2011; Vigiak et al. 2012). Some of those contributions allowed for the identification of bias in data elaboration, due to the utilization of the maximum potentially draining area instead of the actually draining area which is often smaller (Rossi et al. 2015). This allowed for the introduction of a vegetation/land-use effect on the gully-head threshold equation (Torri and Poesen 2014). Rossi et al. (2013), working in another EU-funded project (BIO-SOS), adopted the RECONDES approach, modifying it to deal with data derived by (VH resolution) satellite imageries, and linear artificial structures such as roads, which are major modifiers of water and sediment fluxes. These same concepts were applied for modifying the RUSLE application in catchments, producing a connectivity term which defines the potential sediment contributing area at each position in a catchment (Borselli et al. 2008). The approach was later expanded and applied to mass movement sediment contribution to streams (Borselli et al. 2011) within another EU-funded project (DESIRE).

The aim of the RECONDES project and the research illustrated here was to develop and apply methods of identification of connectivity in the field in order to understand the spatial patterns and dynamics as the basis for development of spatial strategies, and in order to generate data with which to validate models. Because of the lack of methods and examples of actual field mapping of connectivity it was necessary to develop a methods protocol (Cammeraat et al. 2005). Zones and pathways of flow and locations of erosion and sedimentation were mapped from signs of water flows and sediment production/sedimentation that can be visually observed e.g. rills, local deposits, splash pedestals, flow lines delineated simply by alignment of dead leaves. Field mapping of features created in an event required use of a base map with land uses and topographic details. Some mapping prior to events was done to add details of structures etc. and to assess potential (or structural) connectivity. After a rainfall/flow event mapping was undertaken using GPS and photography to mark points. Individual features such as depth of rills and deposits were measured.

To fulfil the objectives, connectivity mapping and assessment needed to be carried out at a variety of scales. This was done mainly at land unit, subcatchment and channel scale but ranged from plot to catchment scale (Fig. 2.1). At each of these hierarchical scales representative areas were selected for detailed mapping and instrumentation. Techniques of mapping needed to vary with scale and type of terrain. The approaches and methods are illustrated in the following sections together with the results on the main patterns, erosion hotspots and pathways. The land units identified are the major ones present in the catchment and typical of upland agricultural areas of the southern Mediterranean region of Europe. These are: afforested land, rainfed croplands and abandoned/semi-natural lands.

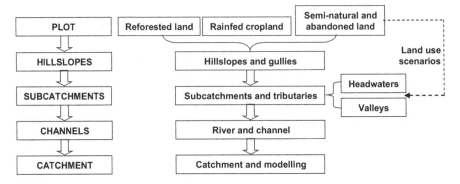

Fig. 2.1 Different scales of analysis of hydrological connectivity and sediment pathways

2.4   Methods and Results at Various Scales

2.4.1   Plot Scale

At the finest scale plot descriptions and maps of vegetation patterns and the relations to the surface, i.e. parent material, crust and erosion features, were made for different runoff response areas (Vijfhuizen 2005). Figure 2.2 gives an example of one of the surveys at the finest scale. This shows the pattern of flows in relation to individual plants and to bare and crusted areas of soil. Mapping and analysis at this scale is relevant if the need is to understand the detailed processes. However, in the overall strategic approach here it is more important to identify the main runoff generating zones or hydrological units and areas vulnerable to erosion and how they are connected downslope.

2.4.2   Land Unit Scale

2.4.2.1   Reforested Land

A field study was conducted to analyze the effects of terracing on the connectivity of a small (1.13 ha) fully reforested catchment (Fig. 2.3). Using field surveys and analysis of orthophoto images with a 0.5 m resolution, five response units were identified. These response units are based on two main criteria: (i) morphology of the unit (slope, type of reforestation/terraces, bedrock, vegetation), and (ii) evolution of erosion and hydrological features. The six units identified were: terraces perpendicular to the main slope, terraces not completely perpendicular to the main slope, steep terraces with long talus, or partially deteriorated, terraces with long natural talus, and sink zones.

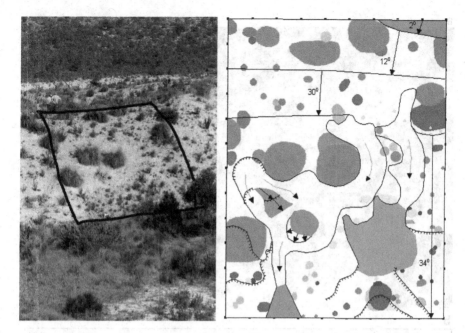

Fig. 2.2 Example of fine scale connectivity mapping. *Left* the actual area and *right* the mapping. *Green dots* indicate different types of vegetation, *arrows* active flow lines of water and lines with hatches small rims, delimiting small terracettes, depressions or channels on the surface (Vijfhuizen 2005)

A map of hydrological connectivity was drawn from the tracks of flow lines by Boix-Fayos, Navarro-Cano and Castillo (Fig. 2.3). It has been made following the guidelines of Cammeraat et al. (2005) (Connectivity and Response Units mapping). The units with less signs of concentrated flow are the *perpendicular terraces* and the *terraces with long natural talus*. The *steep terraces, long talus* unit favours the convergence of flow due to a combination of steep slope and the not complete perpendicularity of the terraces. In other cases, just due to the steep slope, it favours the connection between two terraces breaking the talus through a concentrated flow line. The *terraces not completely perpendicular* unit is the one where more concentrated flow lines appear due to the channelization of water along the main longitudinal slope. In some cases clearly defined long rills and gullies appear parallel to the contour of the terraces, and parallel to the main drainage line of the catchment. Therefore hotspots include defective terraces (not perpendicular to hillslope), localised degraded areas in terraces and semi-collapsed terraces. Steep side banks are also points where erosion and connectivity is favoured.

A total of 40 wooden stakes with water sensitive tape glued on the upslope side were placed all around the catchment in order to test the connectivity map. The stakes were geo-referenced with high-precision GPS. After the storm event on November 2006, water marks on stakes were recorded. From a preliminary analysis of these records, it can be concluded that defective terraces, which are not

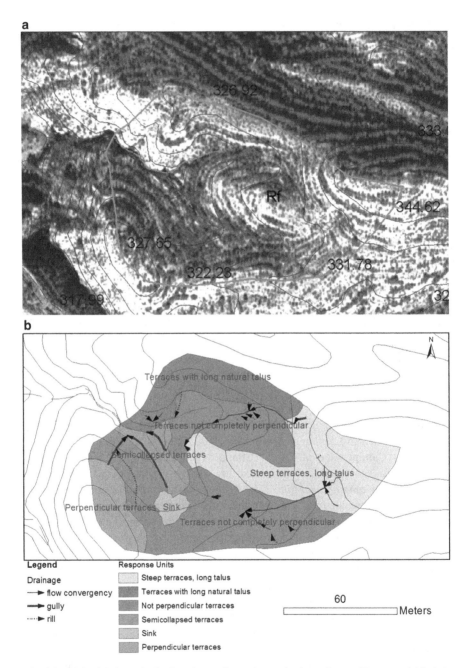

Fig. 2.3 (**a**) Aerial photograph of study area in a reforested sub-catchment (Barranco del Lobo); (**b**) Map of hydrological connectivity in the area: It was identified that reforestation terraces non-perpendicular to the slope increased flow convergence and therefore the initiation of concentrated erosion in the form of rills or gullies (By Boix-Fayos C., Navarro-Cano J. and Castillo, V., in Hooke and Sandercock 2012)

perpendicular to slope, act as fast runoff pathways. Local default in terraces seems to increase the hydrological connectivity. Finally, higher connectivity is favoring the collapse of old reforestation terraces and the migration upstream of the drainage network. This validation confirmed the identified hotspots.

2.4.2.2   Rainfed Cropland

Land degradation by water erosion on rainfed croplands is directly related to the low canopy cover of cropping systems in dry environments. On cereal fields, the application of a fallow year is common, leaving the soil bare throughout the year. In the almond and olive orchards the fields are ploughed several times a year to keep them free of weeds. The aim in this cropland land unit was to assess the feasibility of use of a cover crop to provide a vegetative cover and reduce erosion, particularly under orchard crops. For an effective implementation of cover crops it is necessary to identify the locations in the landscape which promote the connectivity of water and sediment. Such hotspots should have the highest priority in any implementation scheme. Surveys took place after rainfall of different magnitude and intensity in 2006 (Fig. 2.4). The mapping included patterns of concentrated flow and erosion, terrace breaches and pipes.

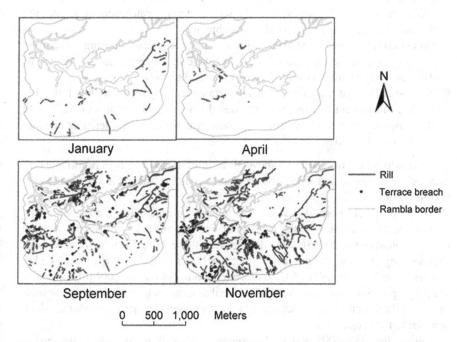

Fig. 2.4 Pathways of concentrated flow in the southern sub-catchment of Cárcavo after rain events in 2006. The mapped terrace breaches include both existing and new breaches (After Meerkerk et al. 2009)

Patterns of runoff, evidence of flow lines and signs of erosion were mapped in the four main rainfall events (Fig. 2.4). Some general observations were made during the field surveys (Meerkerk et al. 2009):

1. There are no runoff/erosion problems on well-maintained terraces with contour tillage. However, on terraces that are longer in the slope direction (e.g. > 20 m) and have a more than gentle slope (e.g. >2°), runoff/erosion is common.
2. The weak parts of the (well maintained) terraces are the access tracks that pass at the sides. Most of these tracks are at the sides or in the middle of the terrace. Sometimes, the tracks continue straight down the hill, passing a large number of terraces. In this configuration, the track forms a channel of concentrated flow. The compacted surface has a low infiltration capacity and will produce runoff for rains as small as 8 mm. Sedimentation occurs in places where the slope is more gentle, at the lower end of the road, and sideways onto the terraces.
3. The shape of the terrace surface is quite important. If a terrace surface has a thalweg, then the probability of concentrated flow increases considerably. An effective/ideal terrace has a straight, horizontal surface perpendicular to the slope.
4. Dirt roads have the highest runoff coefficient of all surfaces and it is here that the first runoff is produced during an event. See for example the connectivity map of January in Fig. 2.4, where runoff is concentrated on road segments. Motha et al. (2004), in a catchment of comparable size in Australia, showed that the road surface may contribute 40–50 % of the sediment at the catchment outlet, despite the fact that it covers only 1 % of the catchment.
5. On agricultural fields erosion/deposition features are masked during most of the year because of the practice of frequent tillage in the orchards and vineyards. They are only visible between the rain event and the next tillage pass.
6. Just a small part of the sediment that was transported by overland flow following the large rain event of November reached the main stream channel (rambla). In the agricultural area, most of the sediment was deposited on terraces or other relatively flat locations.

Summarising the observations above, three important hotspots of runoff and erosion can be distinguished in rainfed croplands: road surfaces, thalwegs and valley bottoms within the fields, and terrace access tracks.

The role of field borders was also examined. An assessment was made of the occurrence and condition of the terraces and retention banks in Cárcavo in 1956 (aerial photos) and in 2005 (fieldwork). Between these dates, the number of terraces decreased by 36 % and the number of retention banks by 28 %. Of the terraces 54 % were intact in 2005 as well as 50 % of the retention banks (Bellin 2006). The remaining part was classified as "permeable", defined as terraces/banks having one or multiple breaches; lacking an intact rim of + 30 cm height; or lacking a counter slope near the terrace edge.

During the 1956–2005 period, the average upslope drainage area of the terrace banks increased from 0.24 to 0.64 ha. These results indicate a clear increase for the potential connectivity within the landscape. The decline in soil conservation

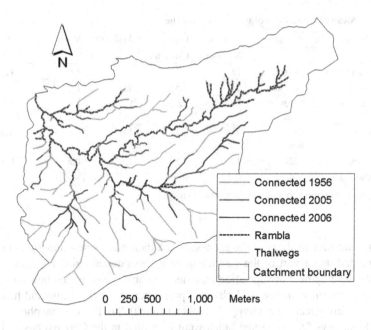

Fig. 2.5   The effect of the decline of conservation structures on the length of the thalwegs that are directly connected to the rambla (After Bellin et al. 2009)

structures is reflected in larger fields and a lower capacity to counteract concentrated flow towards the ramblas (main channels). The observed changes are probably related to the mechanisation of agriculture after 1956. The effect of the decrease of terraces and dams on the connectivity of thalwegs towards the rambla is illustrated in Fig. 2.5.

2.4.2.3   Abandoned Land and Semi-Natural Areas

Similarly, connectivity was mapped in areas that had formerly been used for agriculture. A gully and terrace failure survey was undertaken. In June 2005 abandoned sites and similar cultivated sites were surveyed for gullies to test the hypothesis that abandoned land is more vulnerable to gully erosion than cultivated land. Six field sites were selected, three sets of paired abandoned and cultivated sites, with same lithology and topographic position. The following characteristics were included: gully activity (three classes), vegetation cover in the gully (three classes), type of gully head, slope, and size of the gully. Table 2.1 shows the results of this gully survey. All abandoned sites had more gullies and a higher gully density than the comparable cultivated sites. On one of the cultivated sites no gully at all was found. Also, the estimated volume of each gully was larger on abandoned sites. Gully activity was higher for most abandoned sites, while vegetation cover in the gully was not significantly different (Lesschen et al. 2007).

Table 2.1 Mean characteristics of gullies for each site

Site	Land use	Position	Gullies	Density	Activity[a]	Vegetation[a]	Volume
			Number	Gully/ha			m³
1	Abandoned	Plateau	18	1.1	1.7	1.0	12.4
2	Almond	Plateau	5	0.5	0.6	1.2	4.0
3	Abandoned	Channel head	18	12.6	1.8	0.9	10.1
4	Almond	Channel head	7	2.1	1.9	0.7	3.0
5	Abandoned	Valley	11	2.5	1.6	0.8	0.8
6	Almond	Valley	0	0	–	–	–

[a]Activity and vegetation are ranging from low (1) to high (3)

More intensive analysis of the processes and their effects was undertaken in the instrumented upper part of the catchment under abandoned land. The aim was to study the development of vegetation patterns after abandonment and how these patterns affected soil moisture, soil physical properties at the fine scale and how this affected hydrological connectivity (Lesschen et al. 2008a). From aerial photographs and field surveys 58 abandoned fields were identified in the Cárcavo basin. These fields were surveyed and the following properties were described: previous land use, parent material, vegetation, age of abandonment, erosion features, and presence of terraces or earth dams. Using ArcGIS the field boundaries were digitised and by overlaying the map with a DTM the following properties were determined as well: surface, altitude, slope, drainage area and solar radiation. The 58 fields were then classified into abandoned fields with erosion and without erosion. To be classified as abandoned fields with erosion, the field should show signs of at least moderate erosion, being visible as gully, terrace failure or well developed rill. These two groups were compared using the Pearson's chi-square test for binary variables and the *t*-test for continuous variables. For the detailed analysis of sediment delivery rates we selected an abandoned field, which was located in a valley bottom just before a channel incision.

From the 58 abandoned fields 32 were classified as fields with moderate to severe erosion. Table 2.2 summarizes the average properties for abandoned fields with and without erosion and indicates if differences between the two groups are significant based on the Pearson's chi-square and the *t*-test. The significant ($p < 0.05$) differences between the two groups, based on sufficient observations, are cereals as previous land use, presence of terraces, and maximum slope (Lesschen et al. 2008b). That steeper slopes increase the occurrence of erosion is obvious, but the presence of terraces as a risk factor for erosion seems surprising. The presence of terraces implies that the field is located on sloping land and in addition the terrace topography marks steep gradients on the terrace walls, which makes them vulnerable to gully erosion and piping, especially after abandonment. The last significant factor was cereals as previous land use. An explanation might be the negative land management of cereal cultivation with high soil losses (Lasanta et al. 2000), which

Table 2.2 Averaged properties for abandoned fields with and without erosion

Properties		No erosion	Erosion	Significance[a]
	Number of fields	26	32	
Previous land use (number of fields)				
	Cereals	8	19	**0.030**
	Almonds	14	13	0.315
	Grapes	4	0	*0.021*
Parent material (number of fields)				
	Marl	11	16	0.559
	Colluvium	6	13	0.157
	Keuper	4	2	*0.256*
	Sandstone	3	1	*0.209*
	Limestone	2	0	*0.110*
Vegetation (number of fields)				
	Mainly herbs	12	12	0.506
	Mixed herbs, grasses and shrubs	9	18	0.100
	Mainly shrubs	5	2	*0.131*
Field properties				
	Surface (ha)	2.6	2.0	0.450
	Age of abandonment (year)	9.9	9.2	0.712
	Fields with terraces	7	21	**0.003**
	Fields with earth dams	7	10	0.719
	Mean altitude (m)	389	373	0.230
	Maximum slope (degrees)	13.8	18.1	**0.016**
	Mean slope (degrees)	5.1	5.7	0.288
	Maximum drainage area (ha)	15.3	13.5	0.829
	Mean solar radiation (MJ cm^{-2} year^{-1})	0.82	0.81	0.769

[a]Variables with values in italic had too few observations to be reliably significant

degraded the soil already before abandonment. Figure 2.6 shows the setting of the field site and the calculated sediment losses after subtracting the current DTM with the terrace failures from the 1984 DTM. The main terrace failures are clearly visible with incisions of more than one metre, nevertheless some sedimentation has also occurred, especially below the terrace walls. The average surface lowering since abandonment was 13.8 cm, resulting in a net erosion rate of 87 t ha^{-1} year^{-1}.

This rate is higher than the average 12 t ha^{-1} year^{-1} calculated from severe gully erosion studies (Poesen et al. 2003) and much higher than the usual range of 0.1–1 t ha^{-1} year^{-1} under semi-natural vegetation (Martínez-Fernández and Esteve 2005b). Abandonment of agricultural land is widespread and increasing in Spain and can potentially lead to a considerable increase in erosion in semi-arid areas. Abandoned terrace fields are especially vulnerable because of gully erosion through the terrace walls. In the Cárcavo basin more than half of the abandoned fields have moderate to severe erosion and the calculated sediment delivery rates are high.

Fig. 2.6 Sediment losses for the terrace field since abandonment in 1984

2.4.3   Sub-Catchment Scale

Despite increasing research on connectivity at plot and land unit/hillslope scales, more information is needed about the actual pathways of sediment movement, the position of sources and stores and the influence of spatial arrangement of land uses. An experiment was set up to characterize and assess connectivity for events at the subcatchment scale ($0.1-1$ km^2) under different land use scenarios. This experiment links the research done at land unit/hillslope scale and river channel scale, filling the gap of knowledge between them.

The objectives of this research were to: 1. map the actual pathways of flow and sediment under different land use scenarios for a range of rainfall events; 2. quantify the frequency of response for the identified pathways; 3. estimate rainfall thresholds at this scale; and 4. evaluate the relative weight of relevant factors (land use patterns, topography, geology, roads and tracks, agricultural practices) in influencing connectivity and the delivery of sediment. This analysis also aimed to identify erosion hotspots at the subcatchment scale.

A set of three sub-catchments between 10 and 49 ha in size were chosen within the Cárcavo catchment for detailed connectivity mapping (UOP subcatchments) (Table 2.3).

Table 2.3 Characterization of the studied subcatchments

Id_Exp	Geology	Land use in headwaters	Land use in valley floor	Area (ha)
UoP1	Marls	Abandoned crops	Almond & Olive Trees	14
UoP2_T1	Marls	Reforestation	Reforestation	10
UoP2_T2	Marls	Reforestation	Reforestation	21
UoP2_T3	Marls	Reforestation	Almond & Olive Trees	23
UoP3 (lower)	Marls	Reforestation	Almond & Olive Trees	49

Figure 2.7 shows the results of the mapping and analysis of sources, linkages and sinks (geomorphological map), frequencies of response and rainfall thresholds in the sub-catchment UOP1. The following results emerge:

1. Field configuration (mainly terraces and embankments) restrict flow and sediment movement across the landscape. However, in the largest event some of the embankments were overflowed and broken, so that flow lines tended to follow the main drainage pattern.
2. A great percentage of identified linkages are constant pathways that respond after most rainfall events in spite of constant human labouring and levelling activities.
3. Events of approximately the same size (20 mm) and intensity, as the ones in April 2006 and September 2006, cause different connectivity responses. This may be due to differences in rainfall prior to events.
4. An event occurring after a series of rains, as in November 2006, may cause intense activity in terms of erosion and sedimentation.
5. Sources that responded more frequently were located in the upper part of the subcatchment and characterized by low cover, semi-natural vegetation over shallow and stony soils.
6. The agricultural track concentrated runoff allowing the development of an almost permanent rill that fed the head of the developing piping system in the lower section of the subcatchment.
7. Some rills and features required at least 30 mm of rainfall preceded by over 100 mm in 3 days to show activity. These rills were located mainly in the lower agricultural part of the subcatchment and are frequently ploughed (4–5 times every year).

Analysis of frequency for the recorded activity at UoP3 showed that the most active links (75–100 % frequency) came from southern faced micro-catchments affected by tracks or reforested headwaters. Less active links (0–25 %) were those in the northern-faced vegetated slopes, often having vegetated channels.

The main conclusions of the work exemplified for small catchments are summarised below:

• Repeated mapping of connectivity after rainfall events has been effective in identifying the flow and sediment sources, links and sinks and the frequencies and thresholds of response.

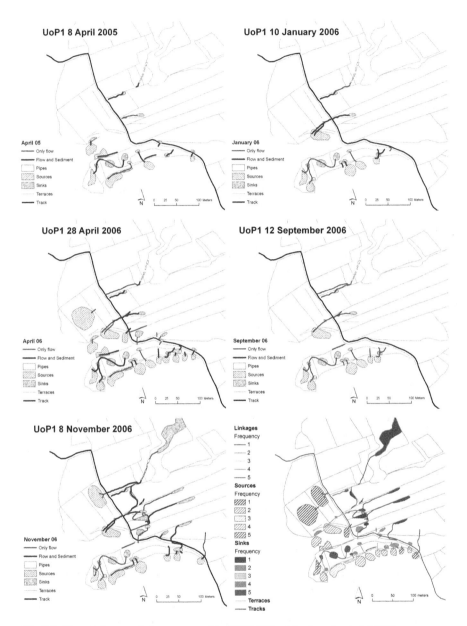

Fig. 2.7 Repeated connectivity assessment in UOP1 (After Marchamalo et al. 2016)

- Connectivity pathways differ in their type of (flow vs. flow & sediment linkages) and in their frequency of response (constant vs. ephemeral). We can distinguish established vs. ephemeral pathways, some of them being event-created pathways and others human-induced pathways (drainage lines, tractor passes).

- Main sources of flow and sediment were identified: these are hardened areas, bare patches, reforested headwaters, low cover south faced slopes and areas affected by roads and tracks. Less active areas were characterized by semi-natural vegetation or the combination of reforestation and semi-natural vegetation in the headwaters and crops in the valley floors.
- Antecedent rainfall was important in inducing greater amounts of runoff and erosion in a 30 mm event in November 2006 than recorded for previous events. This response was characterized by the formation of new rills over the ploughed fields and the activation of previously inactive linkages, which resulted in high overall connectivity for the subcatchments and the whole Cárcavo catchment. During this event the main channel flowed from the headwaters to the lower parts. This formed the basis for establishing thresholds between linkages that reacted for smaller events, those that reacted for November 2006 and those that never reacted during the monitoring period.
- This methodology is useful for identifying hotspots in the landscape where erosion and sedimentation is most likely to occur and connectivity pathways are more frequent. These then form the areas where establishment of vegetation should be encouraged.

Using the evidence from the rainfed cropland areas and the abandoned land the functionality of terraces and vegetation barriers in accumulating water and sediment was further assessed in a sub-catchment approach, combining the flow connectivity observations of Meerkerk et al. (2009) in a catchment with mixed land use, of approx.10 km^2 which forms one of the tributaries of the Carcavo basin. Cumulative runoff, erosion and sedimentation were studied, comparing the current situation with presence or absence of vegetation patterns and terraces. The model shows that the spatial arrangement of sources and sinks at the fine to broader scale (e.g. vegetation patterns or terraces) dictate much of the hydrological (dis)connectivity at the catchment scale (Figs. 2.8 and 2.9) (Lesschen et al. 2009). Rates of sedimentation, erosion and sediment yield at the catchment scale were also calculated and showed that the vegetation and terraces are highly effective in reducing the sediment yield at the broader scale and increasing sedimentation at the local scale (see Table 2.4).

2.4.4   Channels

Processes of degradation along river channels are typically associated with floods. The geomorphic effectiveness of a flood may vary in response to a range of factors including the magnitude, frequency and ordering of previous events and the cumulative effect they have on the channel form (Hooke 2015; Poesen and Hooke 1999; Wolman and Gerson 1978). Thresholds for degradation will vary in association with changes in channel morphology, vegetation type and the calibre of sediments comprising the bed and banks.

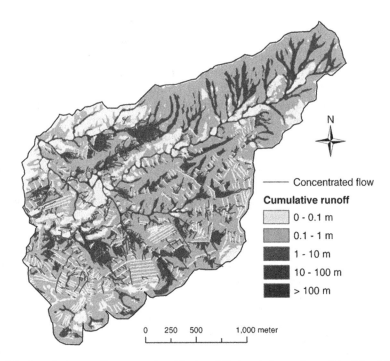

Fig. 2.8 Cumulative runoff concentration (*shades of blue*) and observed runoff (*red lines*) (From Lesschen et al. 2009)

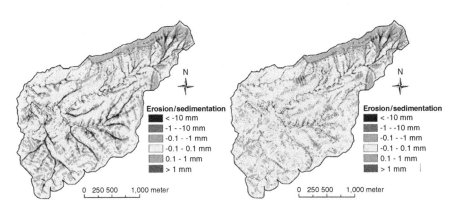

Fig. 2.9 Erosion and sedimentation without terraces (*left*) and with all terraces (*right*) functioning (From Lesschen et al. 2009)

Table 2.4 Model output showing the effect of vegetation and terraces on the erosion and sedimentation (Data after Lesschen et al. 2009)

Processes		Scenario			
		Current	No vegetation	No terraces	All terraces restored
Erosion	Ton ha^{-1}	38.1	56.4	84.8	32.0
Runoff percentage	%	6.8	9.8	25.9	5.2
Sedimentation	Ton ha^{-1}	35.7	49.1	61.7	29.1
Sediment yield	Ton ha^{-1}	2.5	7.4	23.4	3.0

Questions about connectivity of sediment transfers, the nature and distribution of erosion hotspots along channels are addressed in part through channel connectivity mapping, using the method described by Hooke (2003). This method is based on the interpretation of various morphological and sedimentological evidence. In this, the morphology of the channel is mapped, detailing variations in the dimensions of the channel bed and floodplain, bedrock exposures and the position of the check dam/groyne structures and road crossings. Particular attention is given to mapping the sources (gully/tributary, banks/valley walls) and storages of sediment (bar forms, accumulations behind check dams) along the channel. Using the combined map layers (channel morphology, sediment sources and storages) sediment source zones are identified and the channel is divided into areas of erosion, erosion and storage, transfer and net storage.

Check dams have been constructed throughout the Cárcavo catchment in an effort to reduce the erosion and transfer of sediments from incoming tributaries and along the main channel (Castillo et al. 2007). These check dams represent major breaks in the potential connectivity of sediment transfers through the catchment, with large amounts of coarse and fine-grained sediments stored immediately upstream of these structures. They also have a profound effect on patterns of erosion, transport and storage of sediment along the channel. Ten check dams were mapped along the main channel, in varying stages of infill. Road crossings over channels can also act as barriers to sediment transfers, having an effect similar to check dams.

The mapping and interpretation is exemplified in Fig. 2.10 for the upper part of Cárcavo channel. The channel has been divided into segments or reaches. Generalised patterns of potential connectivity can be described for Segments 3–8. A zone of erosion exists downstream of check dams. Within these reaches there is a reduction in the potential for sediment storage, this in part due to the trapping of sediments by the structure upstream and increased erosivity of flows immediately downstream, a product of the lack of sediment load. The channel immediately downstream from the check dam is often scoured to bedrock and there is a notable lack of sediments. Field observations of changes in the cross-sectional shape of the stream channel, the composition of channel bed material, and bankfull stage measurements indicated that the dams cause erosion downstream (Castillo et al. 2007). Where the channel traverses hardened marls, the channel is confined and entrenched to bedrock. Localised areas of scour result in patches of potential ponding that are

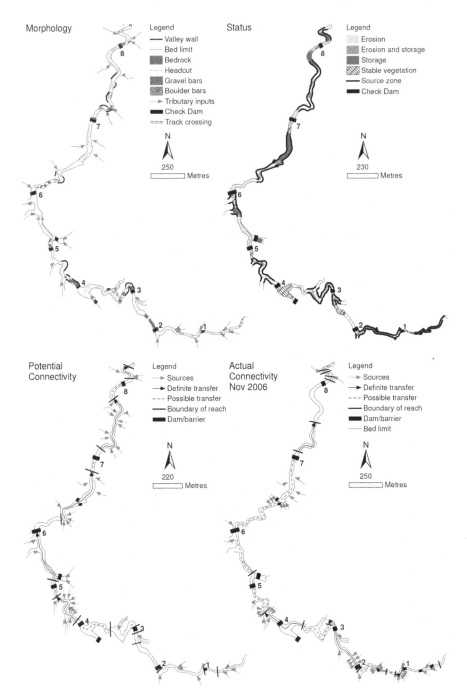

Fig. 2.10 Mapping and interpretation of morphology, status, potential connectivity and actual connectivity along the main channel of Cárcavo for the November 2006 event, upper Cárcavo, Southeast Spain (After Sandercock and Hooke 2011)

characterised by sparse reeds and grasses, which are likely to be frequently removed by floods and sediments flushed through. These reaches are classified as potentially highly connected.

A zone of aggradation extends upstream of the check dam, a response to the reduction in channel gradient. Coarse and fine-grained sediments accumulate upstream of these structures, forming temporary sediment storages. The accumulation of sediments and ponding of water behind these structures provide favourable conditions for the establishment of *Tamarix canariensis* and *Phragmites australis*. As a number of these check dams are completely filled with sediments, it is likely that there are sediment transfers beyond these barriers, however this needed to be confirmed by mapping of actual connectivity for an event.

The lower part of the Cárcavo channel lies more deeply entrenched within Quaternary alluvium and Marls. Based on mapping of sediment sources, it is clear that the majority of coarser material (gravels and boulders) in that part of the channel is being sourced from adjacent valley walls and not from incoming tributaries. Gravel bars form in sections where the channel is supplied with sufficient material from valley walls through mass failure mechanisms and direct erosion of walls by floods.

The 8 November rainfall event of 40 mm, generated sufficient runoff for flow to be recorded in the main channel. Peak stage varied from 0.47 to 0.76 m at different locations, with calculated flood discharge of 1.39–1.94 m^3 s^{-1} as calculated using WinXSPRO software. In the upper parts of the network where the channel has a simple rectangular-shaped morphology, the flow filled the channel floor and had an average depth of 0.3–0.4 m. In the lower reaches the channel has a more compound form, with a smaller inner and a larger outer channel. Flow filled the inner channel to an average depth of 0.5 m. The event had minimal impact on channel morphology and vegetation.

The event was of sufficient magnitude for transfers of sediment to extend beyond a number of check dams. Some tributaries did not contribute sediments to the channel. Of the 36 tributaries which adjoin the main Cárcavo channel, 15 (42 %) did not contribute sediments, whereas the remaining 21 tributaries (58 %) did contribute sediments to the channel. Those tributaries which did not contribute sediments had significantly larger drainage areas (Mean 82 ha) than those contributing sediments (Mean 24 ha). Connectivity mapping has highlighted that sediment inputs to the main channel from incoming tributaries and gullies are actually very low for the events documented. This is not surprising given the high number of check dams that are present throughout the catchment, but it was also not a high flow event. This is also supported by the connectivity mapping done at the land unit and subcatchment scale, which showed that, while there may be significant rilling in fields, the majority of eroded sediments are deposited in the agricultural landscapes at a point not far from the sediment source (within c.50–200 m). Many of the non-contributing tributaries were also highly vegetated in their lower ends. A large proportion of sediments transported along the channel are input directly from valley side walls. In larger catchments more opportunity arises for deposition, in the wider valleys. Also, in this type of environment only large rainfall events generate enough runoff to sustain connected flow through the catchment.

The following areas are identified as erosion hotspots in channels: incoming gullies; confluences; thalweg; areas downstream from check dams; steep valley walls.

2.4.5 Catchment

A method of assessing theoretical connectivity at the catchment scale for application within the modelling environment has been developed by Borselli et al. (2008). This section provides documentation on the equations used, theoretical connectivity maps and validation of these maps in the field. More detailed documentation on the application of theoretical connectivity to modelling is given in Borselli et al. (2008). Various applications and variants can be found in literature after 2010. (For a review please see: http://www.lorenzo-borselli.eu/presentations/Connectivity-SSOG-2015-Borselli.pdf).

The basic idea is that a given spot can be problematic for a series of reasons: (1) it is cut from upslope runoff and consequently does not receive enough water for plants to survive (water stress); (2) it is an excessive sediment/runoff source and contributes substantially to downslope problems (e.g., reservoir siltation, soil erosion). These problems can be managed, at least to a certain extent, if those hotspots are identified and the connection pathway is defined. This allows us to introduce the concept of 'effective catchment' which is the part of the catchment that is really connected to the particular hotspot (which can be the main drainage system as a well as a given field or part of it).

Reasons for disconnection are: (1) ditches, (2) segments of high infiltration rate (e.g. good vegetal cover); (3) area with rough surfaces (high storage capacity); (4) countersloping or large and scarcely sloping areas. The same factors also act for catching transported sediment. To them we can add: (5) any area with high Manning roughness (e.g. grass strips). In most cases of mitigating land degradation the aim is to decrease the connectivity and add connectivity breaks. Obviously, these connection-breakers have less chance to cut a flux the weaker the flux is and the longer the distance the material has to travel to get to a given spot. On this basis (and other more detailed and rigorous scientific bases) the following ratio was proposed as an index of connectivity of a given spot to another spot which is n steps downslope (one step = one cell of size d, gradient s_j and connectivity breaker factor F_j) while the up-slope catchment has an area A_{up} with mean slope $\bar{s}$ and mean connectivity breaker factor $\bar{F}$:

$$I = \log_{10} \left[\frac{\bar{s}\,\bar{F}\sqrt{A_{up}}}{\displaystyle\sum_{\substack{\text{over all downslope} \\ j \text{ segments}}} \dfrac{d_j}{s_j\,F_j}} \right] \tag{2.1}$$

Where:

$$F_j = CNi_j C_j = CNi_j * W_j \qquad (2.2)$$

and where:

W_j is the general weighting factor that in Borselli et al. (2008) is equivalent to C factor in USLE-RUSLE models;

CNi_j is the SCS-curve number method:

Cni, $i = I$, II, III following antecedent moisture conditions.

F_j has been identified with the local curve number value (SCS-curve number method: Cni, $i = I$, II, III following antecedent moisture conditions) when dealing with runoff; and with USLE crop cover factor C. The product $F_j = Cni\ C$ has been used as sediment is connected only if water is connected. The Eq. 2.1 is a variant of the general formula proposed by Borselli et al. (2008) and in this case more oriented to both general runoff generation and potential water erosion. Examples of application to the Cárcavo basin are shown in Figs. 2.11 and 2.12.

The theoretical connectivity produced by the model required validation as suggested also by Borselli et al. (2008) An observation network was established at the end of summer 2006 (described under Reforested land, Sect. 2.4.2.1). A total of 58 plots were established, stratified by the connectivity value of the pixel. Some

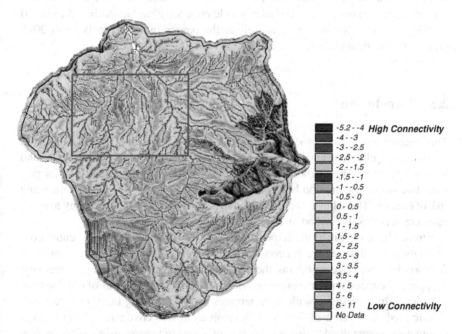

Fig. 2.11 Modelled connectivity (IC Index) for the whole Cárcavo catchment. The connectivity factor here is the product $F_j = Cni\ C$; The *rectangle* indicates the part of the catchment shown in Fig. 2.12

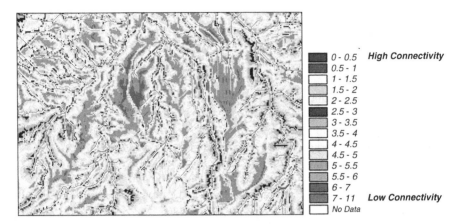

Fig. 2.12 Map of modelled sediment connectivity (IC index) for part of Cárcavo catchment shown in *rectangle* in Fig. 2.11

differences in amount of erosion and deposition were found between sites of different theoretical connectivity, with the very high connectivity sites particularly having greatest changes. In many of the papers now in literature the connectivity Index IC was geomorphologically validated in a variety of environments. (A full literature list can be found also at https://scholar.google.es/scholar?oi=bibs&hl=en&cites=80 62253390937599926&as_sdt=5 as well as the papers indicated in Borselli 2015 presentation indicated above.)

2.5   Conclusions

This chapter has outlined the approach to assessing connectivity and exemplified the detailed techniques and the results at a range of scales and in a variety of land uses and settings. The methods developed enabled the identification of major patterns and pathways, erosion hotspots and factors contributing to vulnerability and risk of erosion. These have then been used in developing the spatial strategy to minimize connectivity, described in Chap. 5.

Studies have identified the important role of terraces in reducing potential connectivity, but that if they are not constructed correctly and maintained they may then form a point where erosion begins, their collapse then further enhancing connectivity. Mapping of connectivity in Reforested Lands highlighted that areas of erosion have a tendency to correspond with steep terraces that have a long talus and are partly deteriorated. Concentrated flow lines develop along defective terraces that are not completely perpendicular. Repeat studies of Cropland areas using aerial photographs (1956) and field surveys (2005) has shown that there have been significant declines in the number of terraces (36 %) and retention banks (28 %), with 54 % of terraces and 50% of retention banks intact in 2005. These results highlight a clear increase in potential connectivity for these areas, with declines in conservation structures reflecting larger fields and probably related to the mechanisation of agriculture during the 50 year period.

Repeat connectivity mapping of subcatchments for a number of rainfall events has shown that many of the same erosion pathways are activated by successive events. Access ramps, tracks and dirt roads represent major source areas for sediments and hotspots where erosion occurs. Surprisingly, whilst significant erosion and runoff resulted from the November 2006 event, very little sediment eroded by overland flow reached the rambla (channel), most of the sediment being deposited on terraces or other relatively flat locations. Repeat checking of gullies/tributaries entering the channel after rainfall events has highlighted that sediment inputs from incoming gullies and tributaries are actually very low, partly a reflection of the high number of check dams but also of vegetation within some sections of channels. A large proportion of the sediments transported along the channel are input directly from valley side walls. Check dams have an overwhelming influence on channel connectivity, these forming major breaks for sediment transfers through the catchment. Significant erosion occurs downstream of these structures due to clearwater effects. However, over time these checkdams fill and thereafter fine sediment is transported but the check dam can act as a 'waterfall', enhancing erosion downstream and danger of collapse. The conclusions on pathways and hotspots are used in development of strategies in Chap. 5.

Chapter 3
Conditions for Growth of Plants

Peter Sandercock, Janet Hooke, Gonzalo Barberá, J.A. Navarro-Cano,
J.I. Querejeta, Jan Peter Lesschen, L.H. Cammeraat, André Meerkerk,
Bas van Wesemael, Sarah De Baets, and Jean Poesen

Abstract This chapter sets out the approach and research methods used to assess
the plant types and species that grow in different parts of the targeted Mediterranean
landscape and that could potentially be used in restoration strategies and mitigation
of desertified and degraded land. Species occurring in the various land units in the
study catchment in southeast Spain are identified. These units are reforested land,
rainfed croplands, semi-natural and abandoned land and stream channels, Factors
restricting growth of trees and understorey vegetation in reforested land were
assessed using experimental plots and the effects of pine litter on seed germination
were tested. The potential for growth of cover crops between orchard trees was
assessed from hydrological balances. Using multivariate statistical analysis various
factors were found to influence the regrowth of vegetation in abandoned agricultural

P. Sandercock (✉)
Jacobs, 80A Mitchell St, PO Box 952, Bendigo, VIC, Australia
e-mail: Peter.Sandercock@jacobs.com

J. Hooke
Department of Geography and Planning, School of Environmental Sciences,
University of Liverpool, Roxby Building, L69 7ZT Liverpool, UK

G. Barberá • J.I. Querejeta
Centro de Edafologia y Biologia Aplicada del Segura (CEBAS),
Department of Soil and Water Conservation, Campus Universitario
de Espinardo, PO Box 164, 30100 Murcia, Spain

J.A. Navarro-Cano
Centro de Investigaciones sobre Desertificacion (CSIC-UV-GV),
Carretera Moncada – Náquera, Km. 4,5, 46113 Moncada, Valencia, Spain

J.P. Lesschen
Alterra, Wageningen University and Research Centre,
Droevendaalsesteeg 3 (GAIA), PO Box 47, 6700AA Wageningen, The Netherlands

L.H. Cammeraat
Instituut voor Biodiversiteit en Ecosysteem Dynamica (IBED) Earth Surface Science,
Universiteit van Amsterdam, Science Park 904, 1098 XH Amsterdam, The Netherlands

© Springer International Publishing Switzerland 2017
J. Hooke, P. Sandercock, *Combating Desertification and Land Degradation*,
SpringerBriefs in Environmental Science, DOI 10.1007/978-3-319-44451-2_3

lands. In the ephemeral stream channels a wide range of variables was analyzed and presence of species found to be clearly related to substrate and hydrological zone. The results on differential conditions necessary for or favouring growth of various species are used in subsequent design of optimal spatial strategies of planting and restoration.

Keywords Reforestation conditions • Pine litter • Cover crops • Abandoned land • Ephemeral channels • Species zonation

3.1 Introduction

The basic approach has been to assess what plant species grow in the different parts of the target environment then to select those species in each land unit which, because of their life-traits, have characteristics that may be used as mitigation measures against erosion and desertification. In order to be able to use these plants it is necessary that measurements are taken in the field of the conditions under which those plants are growing successfully, the conditions where growth is marginal and the conditions where the plants are not growing. On the other hand, the Cárcavo catchment can be considered itself an experiment of massive reforestation in semi-arid conditions intended to: (i) reduce erosion; (ii) improve soil quality; and (iii) accelerate vegetation recovery from a degraded state. The effects of this management strategy on the conditions for growth of plants may be tested. The effects of particular plant characteristics e.g. rooting properties, and of vegetation characteristics e.g. density, pattern, on processes have also been assessed and measured, some by experimentation. Measurements and experiments have also been made on the strength and resistance of the plants to water and flow erosion, including measurements of root strength as well as resistance of aerial parts. The conditions for the growth of plants and those most likely to grow in hotspots and positions that reduce connectivity are described in this chapter. The effectiveness of the plants is assessed in Chap. 4.

A. Meerkerk • B. van Wesemael
Earth and Life Institute, Université catholique de Louvain (UCL), Place Louis Pasteur 3, 1348 Louvain-la-Neuve, Belgium

S. De Baets
College of Life and Environmental Sciences, Department of Geography, University of Exeter, Rennes Drive, EX4 4RJ Exeter, UK

J. Poesen
Division of Geography and Tourism, Department of Earth and Environmental Sciences, KU Leuven, Celestijnenlaan 200E, 3001 Heverlee, Belgium

3.2   Types of Plants in Mediterranean Environment and Land Units

Vegetation present in the two main study catchments in Southeast Spain (Torrealvilla and Cárcavo) and also two catchments in Tuscany, Italy (Landola and Scalonca) has been identified. Research focused on vegetation within the main study area of Southeast Spain (Cárcavo). Detailed reviews of the literature were undertaken to compile the existing information available on conditions, requirements and threshold tolerances for each of the plant species identified (Hooke 2007b). Research focused on the following five different plant functional groups: grasses, herbs, reeds, shrubs and trees. An overview of those plants that are present in each of the major land units used in this study is provided below.

The classification of the life form of the species following Raunkiaer's biotypes, [TER, GEO, HEM, CAM, NAN, FAN, LIA] which uses the vegetative form of a plant based on the position of growth point (buds) during adverse times of the year, has been applied in some cases. TER: Therophytes, annuals, survive in form of seeds; GEO: Geophytes, underground buds (usually bulbous, rhizomatous, etc.); HEM: Hemicryptophytes, buds at soil surface level; CAM: Chamaephytes, buds near ground level (buds < 25 cm high); NAN: Nanophanerophytes, buds near ground level (buds 25–75 cm high); FAN: Phanerophytes, trees and large shrubs; LIA: Lianes, plants growing leant against other plants.

3.2.1   Reforested Lands

Cárcavo catchment is a typical example of an eco-hydrological restoration project carried out in Mediterranean lands in the 1950s–1980s decades. The project goal was simultaneously controlling torrential flows by the construction of checkdams on main channels and their tributaries and afforesting/reforesting hillslopes. This is very similar to the projects carried out from the end of the nineteenth century with the major difference of using massive mechanization of reforestation works mainly by terracing in order to supposedly increase water yield to the planted trees (Querejeta et al. 2001). In Cárcavo catchment 25–30 years after the reforestation works the vegetation characteristics changed from a grassland dominated by the alfa grass (*Stipa tenacissima*) to a cover dominated by the planted Aleppo pine (*Pinus halepensis*) works, with alfa grass as subdominant (Table 3.1 vs. Table 3.2). Despite the increase of vegetation and therefore the soil organic carbon concentration, in the case of Cárcavo catchment, the reforestation strategies (especially those on terraces) have not preserved soil from organic carbon losses by active soil erosion processes, due to often high sedimentological connectivity between reforested terraces and channels (Boix-Fayos et al. in press).

Castillo et al. (2001) studied the development, success and effects of the afforestation programmes on the pre-existent and planted vegetation. The survivorship of

Table 3.1 Dominant species on reforested land directly affected by pine plantation

Species	Mean cover (%)	SE	Life type
Pinus halepensis	23.65	2.997	Tree
Stipa tenacissima	14.46	2.367	Perennial grass
Brachypodium retusum	7.31	2.514	Perennial grass
Rosmarinus officinalis	4.38	0.813	Shrub
Anthyllis cytisoides	0.81	0.546	Shrub
Fumana ericoides	0.77	0.424	Chamaephyte
Quercus coccifera	0.77	0.769	Nanophanerophyte
Thymus membranaceus	0.73	0.232	Chamaephyte
Salsola genistoides	0.58	0.319	Shrub
Fumana thymifolia	0.50	0.262	Chamaephyte

Table 3.2 Dominant species on reforested land on control slopes not affected by pine plantation

Species	Mean cover (%)	SE	Life type
Stipa tenacissima	20.16	2.675	Perennial grass
Rosmarinus officinalis	5.16	1.188	Shrub
Brachypodium retusum	4.74	1.845	Perennial grass
Pinus halepensis	1.68	0.865	Tree
Ononis tridentata	1.21	0.728	Shrub
Anthyllis cytisoides	0.79	0.430	Shrub
Fagonia cretica	0.74	0.737	Chamaephyte
Plantago albicans	0.47	0.300	Hemicyptophyte
Thymus zygis	0.42	0.268	Chamaephyte
Fumana thymifolia	0.37	0.232	Chamaephyte

the planted pine seedlings was only 37 % (CI 95 %, 29–47 %) for reforestations ranging from 5 to >20 years old. It could be expected that survivorship could be negatively related to the time since the restoration if mortality increases with time because interspecific competition is more intense as trees grow, but this was not the case. The only environmental predictor significantly related to survivorship was slope gradient, with higher survivorship on gentler slopes. This result indicates that the intended improvement of water availability by mechanical terracing was not achieved, and is consistent with the increase of hydrological connectivity noted on reforestations with terraces as acting as preferential flow paths (Sect. 2.4.2.1, connectivity on reforested lands). In terms of growth (height and stem diameter of pines) there was a strong variability among plots, in agreement with the very high environmental variation of them in Cárcavo basin, to the extent that the predictive power of statistical models relating growth to environmental variables is low. Specifically, there is a very low correlation between growth and age of the reforestation. This counterintuitive result suggests that harsh conditions severely limit tree growth in Cárcavo, and that the mechanized land preparation (terracing) more than favouring water availability is increasing environmental harshness through increased hydrological connectivity. Tree growth was severely limited by slope gradient, incoming direct solar radiation and the presence of marls and limestone-marls, reinforcing the idea of the negligible positive effect (if any and positive) of mechanical terracing on improving plant growth conditions.

As regards the pre-existing vegetation, the greatest problem is the very poor species richness and vegetation cover on terrace banks. The reasons for this were researched in the framework of the RECONDES project and are reviewed in Sect. 3.3.1.

3.2.2 Rainfed Croplands

In the northern Mediterranean, the most important crops grown under rainfed conditions are olive, grapevine, almond, barley and wheat. The olive (*Olea europaea* L.) is an evergreen tree with a biennial growth cycle that results in one main harvest every 2 years (Ferguson et al. 1994; Rallo and Cuevas 2004; Tubeileh et al. 2004b). The extent of this alternate bearing is greater when tree vigour is low, for example as a result of water and nutrient stress (Ferguson et al. 1994). In contrast to the olive, the almond tree *(Prunus dulcis* Miller*)* is deciduous and does not feature alternate bearing. The major commercial cultivars are self-unfruitful and need cross-pollination to produce almonds (Micke 1996). Therefore, in almond plantations it is common to alternate one row of a major cultivar with one row of a pollinizer cultivar (Micke 1996). Both winter barley (*Hordeum vulgare* L.) and winter wheat (*Triticum aestivum* L.) cereals are cultivated in the northern Mediterranean. They are sown in autumn and harvested before the start of the dry summer. Barley can be grown at drier locations than wheat because it matures 2–3 weeks earlier (Metochis and Orphanos 1997). In this way, barley is less affected by droughts at the end of the growing season and produces higher and more stable yields (Metochis and Orphanos 1997). In spite of the adaptation to drought of almond and barley (the main crops in Cárcavo catchment) their productivity is severely affected by droughts with productivity decrease of more than 50–80 %, respectively, in very dry years (Barberá et al. 1997).

Traditional mixed cropping systems that still occur in the Northern Mediterranean are the dehesa system in West Spain and the montado system in South Portugal (Ceballos and Schnabel 1998; Joffre et al. 1988; Pinto-Correia and Mascarenhas 1999). The montado features a widely spaced tree stand, some 20–80 trees ha^{-1} and a ground cover of arable crops like barley and wheat in rotation with fallow and pasture (Pinto-Correia and Mascarenhas 1999). Typical trees in such systems are cork oak (*Quercus suber* L.), holm oak (*Quercus rotundifolia* L.), some mountain oaks (*Quercus pyrenaica*), olive and sweet chestnut (*Castanea sativa* Miller) (Pinto-Correia and Mascarenhas 1999). Another example of a mixed cropping system is the so called "coltura mista" in Italy, where olive, grapevine and cereals are grown together. The difference with the montado is that this system involves no pasture and livestock, and that the climate is more humid. In modern agriculture, these systems are becoming rare. Instead, new agricultural fields are large and aimed at the production of a single crop. Examples of the intensification and the economy of scale are the modern olive and almond plantations in Spain and the practice of land levelling in Tuscany, Italy (Borselli et al. 2002; Clarke and Rendell 2000; Hooke 2006; Martínez-Casasnovas and Sánchez-Bosch 2000; Ramos and Martínez-Casasnovas 2006).

Cover crops are defined here as crops that are grown to provide soil cover during winter and fallow periods in annual cropping systems and crops that are grown next to the main crop to provide soil cover in perennial cropping systems. The application of cover crops was still rare in the early 2000s in the Mediterranean. Díaz-Ambrona and Mínguez (2001) report that grain legumes cover less than 3 % of the cultivated land in Central Spain. Research by Pastor (2004) demonstrated that it is possible to grow cover crops in rainfed olive plantations during winter. The cover crop consisted of weeds located in between the rows of trees, whereas the soil beneath the trees was kept bare by chemical weeding. When this cover crop is present only during the winter months and is removed in spring, its water use is relatively low (Pastor 2004).

In order to collect more information on the use of cover crops, a field survey was done in Southeast Spain in September 2004, covering parts of the provinces Valencia, Murcia and Jaén. No signs of cover crops were observed, both for non-irrigated and irrigated tree plantations and vineyards. On the other hand, cover crops were observed in Tuscany, Italy (May 2004) and southeast of Grenoble in France (July 2004). It should be noted that the climate at those locations is much more humid than in Southeast Spain, the former having an annual rainfall of 800 mm or more and the latter 300 mm.

3.2.3 Abandoned Lands

After agricultural land abandonment the secondary succession starts and can be described by functional groups, starting with annual or biennial plants and followed by perennial forbs, perennial grasses and shrubs. Annual plants and short-lived perennials have a higher cover and species richness during the first phase of abandonment, forbs during the second phase and woody species increase after 10 years of abandonment (Bonet 2004). Under the semi-arid conditions of Southeast Spain a forest may not be representative of later stages of succession, due to low water resources and intensive human actions in the landscape over millennia. A late successional community composed of shrublands is more plausible for these areas (Rivas-Martínez 1987). As a result of changes in vegetation and soil management the soil properties of abandoned fields will change with time of abandonment. In general, a progressive recovery of vegetation cover, litter production, organic matter, water retention capacity and stability of aggregates takes place on abandoned fields (Bonet 2004; Cammeraat et al. 2010; Cerdà 1997; Dunjo et al. 2003; Kosmas et al. 2000; Martínez-Fernández and Esteve 2005a; Martinez-Fernandez et al. 1995).

Table 3.3 Description of the field sites (From Lesschen et al. 2008a)

ID	Substrate	Stage	Gravel cover (%)	Crust type	Slope (%)	Slope form	Main vegetation species
M1	Marl	Fallow	5	Slaking	4	Straight	Annuals
M2	Marl	±6 years abandoned	5	Slaking	0	Concave	*Bromus* sp., *Eryngium* sp., *Artemisia herba-alba*, *Plantago albicans*
M3	Marl	±25 years abandoned	14	Slaking	0	Concave	*Plantago albicans*, *Artemisia herba-alba*, *Lygeum spartum*
M4	Marl	Semi-natural	13	Cryptogamic	3	Convex	*Quercus coccifera*, *Brachypodium retusum*
C1	Calcrete	Fallow	46	Slaking	5	Convex	Annuals
C2	Calcrete	±9 years abandoned	88	Sieving	1	Convex	*Helichrysum stoechas*, *Artemisia herba-alba*, *Teucrium capitatum*
C3	Calcrete	±40 years abandoned	35	Sieving and cryptogamic	2	Straight	*Stipa tenacissima*, *Thymus* sp., *Teucrium* sp., *Artemisia herba-alba*
C4	Calcrete	Semi-natural	75	Sieving and cryptogamic	15	Straight	*Rosmarinus officinalis*, *Stipa tenacissima*

We studied the development of spatial heterogeneity in vegetation and soil properties after land abandonment on two different lithological substrates in Southeast Spain. Based on the space-time substitution approach (Pain 1985), to overcome the problem of long term monitoring, two sequences of abandoned fields were selected (fallow, recently abandoned, long abandoned and semi-natural vegetation) in the south-eastern part of the Cárcavo basin. One sequence of fields is located on a Cretaceous marl substrate (M-sites) and the other sequence on Quaternary colluvial pediment with a calcrete, which was partly broken by ploughing (C-sites). The main difference between the two substrates is the amount of gravel on the soil, which influences crust formation and infiltration. On these fields a vegetation survey was conducted and soil samples collected. In each field a representative plot of 10×10 m was selected and all species and the estimated cover were recorded, using Sánchez Gómez and Guerra Montes (2003). Afterwards the vegetation species were classified according to the vegetation type, i.e. herbs, grasses, shrubs or trees. Furthermore, a description of each field in terms of topography, crust and vegetation was made (Table 3.3). The age of abandonment was estimated using aerial photographs of 1956, 1985, 1997, 1999 and 2002 and by asking the owner of the fields.

On each field site four topsoil samples (0–5 cm) were taken from bare patches and four from vegetated patches on each field. The following analyses were carried out on the soil samples: organic carbon, aggregate stability, pH and EC. Furthermore, one subsoil sample of each site was analysed for $CaCO_3$ content, soil texture, micro-aggregation, soluble salts and sodium absorption ratio (SAR_p).

For calcrete sites a clear decrease in number of species with time since abandonment was found, while this trend was less pronounced for marl. This decrease is mainly associated with the decrease in annuals and herbs, which for marl is only observed on the semi-natural site. Although the number of shrub species barely increased with time since abandonment, the shrub cover increased considerably. On the fallow sites mainly annual herbs like *Anagallis arvensis*, *Moricandia arvensis*, *Centaurea aspera* and *Filago pyramidata* grew, but also some chamaephytes such as *Artemisia herba-alba* and *Teucrium capitatum* were found, but none of the species was dominant. On the recently and long abandoned sites on marl patches of herbs (*Carrichtera annua*, *Eryngium sp.* and *Eruca vesicaria*), grasses (*Bromus sp.*, *Lygeum spartum* and *Dactylus glomerata*) and few chamaephytes and shrubs (*Artemisia herba-alba* and *Salsola genistoides*) occurred with *Plantago albicans* in between on bare patches. The recently abandoned site on calcrete was dominated by the chamaephytes *Artemisia herba-alba*, *Helichrysum stoechas* and *Teucrium capitatum*. The long abandoned site on calcrete had similar plant species as the semi-natural vegetation, like the perennial bunchgrass *Stipa tenacissima* and the shrub *Rosmarinus officinalis*, but the occurrence of the chamaephyte *Artemisia herba-alba* and some hemicryptophyte *Plantago albicans* indicated its former agricultural use. Finally, on the semi-natural sites mainly *Stipa tenacissima* and *Rosmarinus officinalis* were found with some *Quercus coccifera* shrubs and *Pinus halepensis* trees, and for the site on marl also some *Brachypodium retusum* grass. Vegetation recovery after land abandonment in a semi-arid environment appears, from the field evidence and dating of abandonment, to be slow and to take at least 40 years. This recovery rate is much slower compared to more humid environments in the Mediterranean.

3.2.4   All Land Units

Combining the different land units, the vegetation of 134 plots was described and all species were identified and their cover was estimated.

Table 3.4 shows the distribution of the species over the different land uses and indicates significant differences between abandoned land and semi-natural vegetation and between natural forest and reforestations. The abandoned land can be distinguished from the semi-natural vegetation by the lower occurrence of *Stipa t.*, *Rosmarinus o.*, *Pinus h.*, *Cistus c.* and *Helianthemum sp.* and the higher occurrence of *Atractylis h.*, *Plantago a.* and *Poa a.*. The difference between natural forest and reforestation is made by the lower occurrence of *Pinus h.*, *Brachypodium r.*, *Quercus c.*, *Juniperus o.*, *Lapiedra m.*, *Pistacia l.* and *Cistus a.* on reforested land. The lower occurrence of *Pinus h.* on reforested land seems strange, since all reforestations in the study area consist of *Pinus h.*, however, this may be an indication of the failure

Table 3.4 Species per land use

Species	N	Abandoned	Semi-natural	Natural forest	Reforestation
		%	%	%	%
Stipa t.	119	62[a]	92	82	94
Rosmarinus o.	111	31[a]	84	94	89
Pinus h.	83	0[a]	37	100	79[a]
Thymus v.	77	54	61	53	58
Brachypodium r.	65	31	42	76	48[a]
Cistus c.	57	15[a]	55	53	38
Helianthemum sp.	53	8[a]	39	59	41
Anthyllis c.	51	38	42	29	38
Sedum a.	46	8	34	59	33
Atractylis h.	41	69[a]	34	12	26
Teucrium c.	38	23	11	18	42
Helichrysum d.	36	15	32	24	27
Asparagus h.	33	46	24	6	26
Rhamnus l.	29	15	32	29	15
Quercus c.	29	0	21	59	17[a]
Juniperus o.	27	0	18	65	14[a]
Salsola g.	25	31	16	18	18
Asphodelus f.	25	15	29	18	14
Lapiedra m.	25	0	18	53	14[a]
Pistacia l.	24	15	26	29	11[a]
Juniperus p.	21	8	18	29	12
Stipa c.	20	8	16	29	12
Herniaria f.	16	8	18	0	12
Cistus a.	15	0	8	41	8[a]
Lithodora f.	15	0	21	35	2
Teucrium p.	15	15	11	0	14
Plantago a.	14	46[a]	11	6	5
Diplotaxis l.	13	8	13	12	8
Thymus z.	13	8	13	12	8
Helianthemum a.	12	0	24	6	3
Ononis t.	11	0	3	0	15
Artemisia h.	11	23	13	0	5
Poa a.	10	46[a]	3	0	5
Number of species		52	65	38	63

[a]Significant difference (P<0.05) between abandoned land and semi-natural vegetation or forest and reforestation

of the reforestation program since part of the *Pinus h.* has died. When comparing the total number of species, the reforested areas have many species, which is probably related to the previous land use of semi-natural vegetation, which has the highest species richness. The natural forest has less species, but more late successional species such as *Quercus, Juniperus* and *Pistacia.*

3.2.5   Hillslopes and Gullies

The major focus of studies working at the scale of hillslope and gullies has been on the role of roots in increasing erosion resistance against concentrated flow. Information on root characteristics of Mediterranean plants is limited. Therefore, various characteristics were assessed for 26 typical Mediterranean plant species. Plant species of three different habitats were selected, i.e. phreatophytes or species growing in gullies or channels (Group 1), species growing on Quaternary loam deposits, frequently found on abandoned croplands (Group 2) and species growing in marls on steep badland slopes (Group 3). The species identified in each zone and some of their characteristics are listed in Table 3.5.

The root to shoot (R:S) ratio is generally smaller than 1 (Table 3.5), indicating that the above ground biomass is usually higher than the below ground biomass for the species measured in this study. In most cases, this can be attributed to an underestimation of the total below ground biomass due to the lack of measurements at depths $> d_{max}$ (see Table 3.5). The soil depth at which 50% of the total measured below-ground biomass is present varies between 10 and 20 cm for species growing in concentrated flow zones (group 1) and on steep badland slopes (Group 3). All species growing on relatively flat (slope of 1–2%) abandoned croplands (Group 2) have 50% of their below ground biomass present in the first 10 cm of the soil. The depth at which roots concentrate is most probably linked to the availability of soil water, but the influence of the vertical distribution of available nutrients which are strongly concentrated in surface soil layers, and the need for plant anchorage should not be neglected. For each particular environmental setting the balance between water and nutrient acquisition and the necessity for plant anchorage on unstable geomorphological configurations may mark differences in root development and its vertical distribution. Therefore, species growing in channels have more roots at greater depths compared to species growing on flat abandoned fields. Species from Group 3 (growing on steep badland slopes) have the highest root mass at greater depths. Deeper roots are needed in this habitat to ensure anchorage. Species belonging to group 2 are generally not so deeply rooted (max. 30 cm, except for *Dorycnium pentaphyllum* and *Rosmarinus officinalis*) compared to species from group 1 (ranging from 30 to 90 cm deep). The measured root depth for plant species from group 3 ranges between 20 and 80 cm.

3.2.6   Channels

The types and distribution of plants vary markedly along the channels, but it is possible to identify distinctive variations and patterning. This variation has been the basis for further detailed studies of the conditions under which the plants grow, monitoring of plants and interactions with channel processes. A list of the dominant plants identified within the Cárcavo and Torrealvilla channels is presented in Table 3.6.

Both Cárcavo and Torrealvilla channels share the same major plant species, as listed in Table 3.6. Annual and perennial grasses exist in the channels. Perennial

Table 3.5 Characteristics of 26 plant species in the Cárcavo basin

Name of the species	N	H (cm)	D_{sv} (cm)	d_{max} (cm)	% of total root mass				R:S
					D>2 mm	2<D<5 mm	5<D<10 mm	D>10 mm	
GROUP 1									
Atriplex halimus [Shrub]	5	62	51	60	3	12	7	78	0.70
Dittrichia viscosa [Shrub]	6	47	13	40	2	40	37	21	0.41
Juncus acutus [Reed]	5	40	19	40	55	0	45	0	0.49
Limonium supinum [Herb]	5	45	19	30	39	56	5	0	0.23
Lygeum spartum [Grass]	5	66	24	30	57	0	43	0	0.46
Nerium oleander [Shrub]	2	88	95	50	0	2	11	88	0.86
Phragmites australis [Reed]	5	40	3	40	3	12	81	4	1.02
Retama sphaerocarpa [Shrub]	4	140	51	40	1	5	8	86	0.68
Tamarix canariensis [Tree]	5	59	17	90	10	40	44	6	0.75
GROUP 2									
Artemisia barrelieri [Shrub]	4	34	34	30	43	35	15	7	0.14
Avenula bromoides [Grass]	5	58	11	20	100	0	0	0	0.44
Brachypodium retusum [Grass]	4	28	14	20	100	0	0	0	0.53
Bromus rubens [Grass]	28	8	1	10	100	0	0	0	0.07
Dorycnium pentaphyllum [Shrub]	5	49	29	50	4	18	31	47	0.19
Fumana thymifolia [Shrub]	5	6	9	20	33	67	0	0	0.18
Piptatherum miliaceum [Grass]	5	119	23	30	34	0	66	0	0.16
Plantago albicans [Herb]	18	5	2	30	100	0	0	0	1.50
Rosmarinus officinalis [Shrub]	4	50	34	60	10	39	22	29	0.29
Teucrium capitatum [Shrub]	5	27	20	20	47	29	24	0	0.15
Thymus zygis [Shrub]	5	33	29	30	18	60	14	8	0.22
Thymelaea hirsuta [Shrub]	5	62	38	40	7	27	38	28	0.18

(continued)

Table 3.5 (continued)

Name of the species	N	H (cm)	D_{sv} (cm)	d_{max} (cm)	% of total root mass				R:S
					D>2 mm	2<D<5 mm	5<D<10 mm	D>10 mm	
GROUP 3									
Anthyllis cytisoides [Shrub]	5	80	27	60	4	43	17	36	0.31
Helictotrichon filifolium [Grass]	5	57	16	20	73	0	27	0	0.88
Ononis tridentata [Shrub]	3	39	42	70	3	16	30	51	0.38
Salsola genistoides [Shrub]	5	65	39	80	10	41	17	31	0.23
Stipa tenacissima [Grass]	8	70	51	40	83	0	17	0	0.15

Group 1 are phreatophytes or species growing in channels. Group 2 are species growing on abandoned crop-lands. Group 3 are species growing in marls on steep badland slopes. N is number of selected plants, H (cm) is mean plant height, D_{sv} (cm) is mean diameter of the rooted soil volume, D_{max} (cm) is the maximum sampled root depth, D is root diameter and R:S is the mean root to shoot ratio (g g^{-1}) (After De Baets et al. 2007b)

Table 3.6 Plant functional group and species identified along Cárcavo and Torrealvilla channels in SE Spain

Plant functional group	Ephemeral channel species list	
	Torrealvilla	Cárcavo
Trees	*Eucalyptus (sp)* [FAN]	*Eucalyptus (sp)* [FAN]
	Populus nigra [FAN]	*Pinus halepensis* [FAN]
	Tamarix canariensis [NAN]	*Tamarix canariensis* [NAN]
Shrubs	*Anthyllis cytisoides* [NAN]	*Anthyllis cytisoides [NAN]*
	Artemisia barrelieri [CAM]	*Ballota hirsuta [CAM]*
	Ballota hirsuta [CAM]	*Dittrichia viscosa [CAM]*
	Dittrichia viscosa [CAM]	*Genista spartioides* [NAN]
	Dorycnium pentaphyllum [NAN]	*Nerium oleander* [NAN]
	Genista spartioides [NAN]	*Rosmarinus officinalis [NAN]*
	Nerium oleander [NAN]	*Salsola genistoides [NAN]*
	Retama sphaerocarpa [NAN]	*Suaeda vera [NAN]*
	Rosmarinus officinalis [NAN]	
	Salsola genistoides [NAN]	
	Senecio linifialaster [CAM]	
	Suaeda vera [NAN]	
	Thymelaea hirsuta [NAN]	
Herbs	*Foeniculum vulgare* [HEM]	*Limonium (sp)* [HEM]
	Limonium (sp) [HEM]	
	Polygonum equisetiforme [TER]	
Grasses	*Lygeum spartum [HEM]*	*Lygeum spartum [HEM]*
	Piptatherum miliaceum [HEM]	*Piptatherum miliaceum* [HEM]
Reeds	*Juncus maritimus [GEO]*	*Juncus acutus [GEO]*
	Phragmites australis [GEO]	*Juncus maritimus [GEO]*
	Saccharum ravennae [GEO]	*Phragmites australis* [GEO]
	Typha dominguensis [GEO]	*Saccharum ravennae [GEO]*
		Schoenus nigricans [GEO]
		Scirpus holoschoenus [GEO]
		Scirpus maritimus [GEO]

[TER, GEO, HEM, CAM, NAN, FAN, LIA] refer to the lifeform of the species following Raunkiaer's biotypes, which uses the vegetative form of a plant based on the position of growth point (buds) during adverse times of the year. *TER* Therophytes, annuals, survive in form of seeds; GEO: Geophytes, underground buds (usually bulbous, rhizomatous, etc.); *HEM* Hemicryptophytes, buds at soil surface level; *CAM* Chamaephytes, buds near ground level (buds < 25 cm high); *NAN* Nanophanerophytes, buds near ground level (buds 25–75 cm high); *FAN* Phanerophytes, trees and large shrubs; *LIA* Lianes, plants growing leant against other plants

species are of greater interest for erosional control as they provide a permanent ground cover throughout the year. The perennial *Lygeum spartum* is a tussocky grass with an extensive rooting system, that can form a dense cover over the bed of marly channels (Navarro-Cano 2004). These communities have been identified as playing a role in countering erosion and desertification; the root systems of these tussocks are particularly effective on steep slopes (Garcia-Fuentes et al. 2001;

Mattia et al. 2005). *Piptatherum miliaceum* and *Brachypodium retusum* are also present. Dominant herb and chamaephyte species present include *Foeniculum vulgare, Limonium sp.* and *Polygonum equisetiforme.* Their occurrence within the ramblas is highly opportunistic, often growing where conditions are too hostile for more demanding plants. They have a low resistance to erosion and are easily washed away by floods (Johnson and Simpson 1985), with Mant (2002) documenting their removal by minor to moderate floods.

A range of reed species are present along the channels, each adapted to varying levels of water hydroperiod (the average duration of water saturating conditions in the substrate) and soil salinity. *Juncus maritimus* and *Phragmites australis* have deep growing rhizomatous root structures, that spread laterally, permitting individual plants to quickly form extensive stands (Hara et al. 1993; Snogerup 1993). Other reed species documented include *Juncus acutus, Saccharum ravennae, Scirpus holoschoenus, Scirpus maritimus, Schoenus nigricans* and *Typha dominguensis.*

Dominant shrub species are *Anthyllis cytisoides, Artemisia barrelieri, Ballota hirsuta, Dittrichia viscosa, Dorycnium pentaphyllum, Genista spartioides, Nerium oleander, Retama sphaerocarpa, Rosmarinus officinalis, Salsola genistoides, Senecio linifoliaster, Suaeda vera* and *Thymelaea hirsuta.* Most of the literature pertaining to shrubs in the Mediterranean is based on hillslopes and not channel environments. Moisture and aspect are two factors considered particularly important in influencing the density of shrub seedlings (Smith et al. 1993). In comparison to grasses, herbs and reeds, shrubs contribute a greater roughness value to the channel. Resistance to erosion varies markedly between the different species, with *N. oleander,* noted for its high resistance to erosion (Herrera 1991).

Tamarix canariensis is the dominant tree species found in the channels. A moist substrate for a period of 4–6 weeks is required for seedlings to establish (Brock 1994). They are able to tolerate extended periods of drought, flooding and sedimentation and also have a high salt tolerance (Birkeland 1996). They may form dense stands in close proximity to dams, this being in part due to their higher stress tolerance and ability to reproduce under a range of conditions (Levine and Stromberg 2001). This species has a high resistance to erosion; an extensive search of the literature has not revealed any studies where flows are quoted as having been effective in removing individuals. Other tree species found occasionally near the channels include *Eucalyptus camaldulensis* and *Populus nigra.*

3.3 Assessment of Conditions for Plants

3.3.1 Reforested Lands

In Sect. 3.2.1 the environmental template created by pine reforestation in Cárcavo was briefly reviewed. The reforestation does not only intend to introduce and promote the growth of one or some planted species, as its main underlying premise

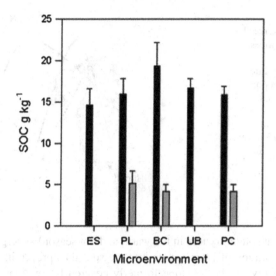

Fig. 3.1 Soil organic carbon on reforestation microenvironments, *ES* under the escarpment in the transition bank/terrace, *PL* adjacent to the stem of pines planted on the terrace, *BC* beneath tree crown cover, *UB* external border of the terrace and upper rim of bank, *PC* pseudocontrol, remaining hillslope vegetation unaltered by mechanical works. *Black bars*, former seminatural vegetation on hillslope; *grey bar*, former barley field. ES and UB microenvironment were absent on former barley fields as they were produced by mechanical terracing of the hillslope. (Source: Modified from Ruiz-Navarro et al. (2009))

hypothesis is that this planted vegetation will increase soil quality and, in general, other environmental conditions favoring faster secondary succession. Cárcavo reforested land was an excellent bench mark to test this hypothesis for traditional approaches to land reforestation in the framework of a harsh environment. In the RECONDES project it was studied whether soil quality improves with reforestation and how the planted vegetation (Aleppo pine) conditions other species. In order to test the effect of the reforestation on soil quality twenty plots were set on one of the oldest reforested areas (about 30 years), ten plots on a terraced hillslope formerly covered by seminatural vegetation and ten on a nearly flat area formerly cultivated with barley (Ruiz-Navarro et al. 2009). The working hypotheses were that tree development would generate changes in the soil (reflected in a radial gradient around planted trees), and that disturbance by mechanical site preparation would result in some microenvironments with unfavorable conditions for plant growth.

On abandoned barley fields, pine litter layer thickness was clearly distributed in a radial gradient around trees. However, despite the large differences in litter abundance, there were no differences in soil organic carbon (SOC) along this gradient, nor with the unaltered soil representing conditions prior to reforestation (Fig. 3.1). Thus, after 30 years of development of the planted vegetation the purpose of improving soil quality was not reached or even approached. The recalcitrant nature of pine litter combined with the unfavorable conditions for decomposition in semi-arid Mediterranean conditions (low average rainfall; moderately wet conditions in the

Fig. 3.2 Soil characteristics on terraced and reforested hillslopes. The sediment flows decouple SOC from litter input spatial distribution. Scale is exaggerated for visualization, true differences in SOC are shown in Fig. 3.1. (Source: Ruiz-Navarro et al. (2009))

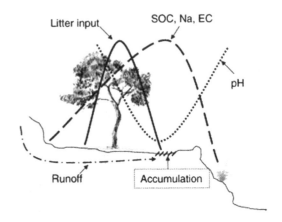

cool winter and absolute dryness in the warm and hot season) synergistically act to delay the incorporation of litter carbon into the soil, thus preventing the improvement of soil quality. In the hillslope formerly covered by seminatural vegetation SOC was three times higher on average than in the former barley fields, which is an indicator of the very poor soil status induced by long barley cultivation (see Sect. 2.4.2.3 for a discussion about the degrading effect of cereal cultivation in the catchment), but not an indication of better effects of reforestation in the hillslope with seminatural vegetation. Differences of SOC between tree influenced microenvironments were minimal compared to pseudocontrols on unaltered hillslopes (Fig. 3.1). In terraced hillslopes the SOC distribution was clearly spatially uncoupled from that of litter layer depth. This was an effect of sediment redistribution by runoff within the terrace that preferentially moved fine sediments to terrace rims (Fig. 3.2).

Additional experiments were set up to elucidate if the establishment of a tree cover may improve the activity of biogeochemical cycles in soils (Bastida et al. 2008). Soil microbiota activity has been usually overlooked but as responsible for nutrient cycling and organic matter decomposition is key in the process of restoring degraded lands. Four experimental plots, two on north facing slopes (75 % vegetation cover) and two on south facing slopes (25 % vegetation cover) representing, respectively, moderately wet vs. dry conditions were established. In each topographical position one plot was covered by pine forest and the other by shrubland. Microbial activity was markedly higher on north facing slopes than on south facing slopes with average respiration rates of 15 mg $CO2$-C kg^{-1} soil^{-1} day^{-1}, vs. less than 2 mg $CO2$-C kg^{-1} soil^{-1} day^{-1}, respectively. This was in agreement with the enzymatic activity involved in biogeochemical cycles, like for example protease activity linked to N cycling. The effect of the type of vegetation cover was weak (if any) and undetectable at the most favourable site, the north facing slope. In summary, the study of the microbiological activity in soils shows that the topographic controls were much stronger than the supposed benefit of a new tree cover over the pre-existing shrubland. This is also in agreement with the very weak improvement of soil quality by afforestation/reforestation in the medium term as observed in the previous experiment.

The pine development in reforestation leads to considerable litter accumulation in comparison to the pre-reforestation situation. As we saw in the previous paragraphs this accumulation does not translate in the medium term (about 30 years) into improved soil quality but it could interfere with seed germination and seedling emergence. Control on plant communities by plant litter has been pointed out on numerous occasions. It is exerted not only by the nutrient levels, moisture, temperature and the allelopathic effects of leaves or other organs (Carson and Peterson 1990; McAlpine and Drake 2002; Xiong and Nilsson 1999) but also by the physical properties of litter (Facelli and S. 1991). These factors may differentially act on germination and seedling establishment (Baskin and Baskin 1998; Facelli 1994; Rice 1979), that represent the most critical period for population dynamics in plants (Harper 1977; Kitajima and Fenner 2000).

Pinus litter layer is present both in natural and planted pinewoods, usually showing a thickness of several centimetres (McAlpine and Drake 2002). Several authors have estimated the mean time necessary for pine needle decomposition as more than 4 years (Moro and Domingo 2000; Prescott et al. 2004) but in Cárcavo conditions it is estimated to be much longer. Allelopathic control has not been detected in the plant dynamic regulation of *P. halepensis* semi-arid plantations (Maestre and Cortina 2004), although Izhaki et al. (2000) and Eshel et al. (2000) pointed out the quimiotoxic effect of the post-fire pine ash on the plant succession of the pine understorey in subhumid Mediterranean conditions. On the contrary, the physical effect of the pine litter could play an important role in seedling emergence of early successional species (Izhaki et al. 2000). During the project RECONDES both field and laboratory experiments were carried out to test several hypotheses about how pine litter could condition seedling emergence and growth.

In the field, direct sowing of *Stipa tenacissima* and *Anthyllis cytisoides*, two dominant species in the shrublands and grasslands and in the pine understorey, was carried out on the footslope site in order to test the hypothesis of a negative interaction between litter accumulation and seedling emergence (Navarro-Cano et al. 2009, for consistency with Ruiz-Navarro et al. 2009 we changed here terms to name microenvironments). Each plot was divided into three microenvironments below pine plantation line (PL) adjacent to the trunk, below pine canopy (BC) and bare soil between plantation lines unaffected by reforestation and acting as pseudo-controls of conditions prior to the reforestation (PC). *S. tenacissima* seedling emergence began 26 days after sowing. Two germination periods took place. In a first period, from 6 April to 25 April, significant differences between microsites were found (One-way ANOVA, $F_{2,14}=7.586$; $P=0.006$). There was a gradual increase in the seedling emergence mean rate from PL ($1\pm0.65\%$) to PC ($40\pm11.9\%$), following a gradient of needle thickness. However, seedling survival was nil in all the microsites 3 months after sowing. We detected new seedling emergence during the following autumn, but all the new plants grew between 28/9/05 and 19/10/05. The accumulated seedling emergence reached $3.5\pm0.9\%$ in PL, $26\pm7.2\%$ in BC and $45\pm11.1\%$ in PC (Fig. 3.3a). Significant differences were only found between PL and PC microsites (Tukey test, $P=0.004$), as happened in the spring period. However, marginally significant differences were found now between BP and BC

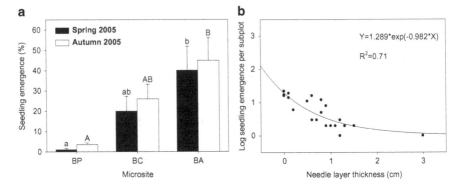

Fig. 3.3 (**a**) Variation of seedling emergence among microsites. Columns marked by different letters indicate significant differences (Tukey, $P<0.05$); (**b**) The relationship between seedling emergence per plot and the needle layer thickness. Seedling emergence was previously $\log(x+1)$ transformed to pass Normality and Constant Variance tests. Fitted line is for exponential decay equation

(Tukey test, $P=0.053$). As a benchmark the germination of *S. tenacissima* under controlled conditions on a growth chamber reached $69.3 \pm 3.4\%$.

The hypothesis that *Pinus* litter layer can act to hinder seedling emergence is supported by the results, which show a decrease of seedling emergence when *Pinus* litter layer increases (Fig. 3.3b). Considering that germination of *S. tenacissima* is not affected by the light environment (Gasque 1999), our results support the hypothesis that the fine-scale distribution of the *P. halepensis* needle layer hinders seedling emergence by a physical effect. Needles form a very intricate and porous layer, which might hamper burial, seed-soil contact and seed imbibition. In this situation, once the seed has germinated, moisture and temperature stress conditions can negatively affect the radicule rooting in the soil and subsequent seedling emergence. Since the negative effect of litter disappeared when water balance was improved in the growth chamber, our results support this interpretation. In any case, we detected a significant decrease of *Stipa* seedling emergence with a mean needle layer thickness of 1.3 cm (PL microsite), but litter layer is usually larger than 2 cm in older pine plantations with a larger size than in our study site (personal observation) so a generalization of this effect may be expected in other semi-arid afforested areas.

In semi-arid ecosystems there are a plethora of species that are early-colonizers capable of occupying highly stressful niches. In the previous experiment we ascertain how pine reforestation can affect the seedling establishment of subdominant understorey species. A second experiment was conducted to investigate how pine litter may affect species that have this role of early-colonizers, and whose ecology may be very different to that of the subdominant species. On the other hand the effect of litter may imply alelopathic effects and the previous experiment was not designed to evaluate this possibility. The herb *Diplotaxis harra* and the chamephytes *Thymus zygis* and *Teucrium capitatum.* were selected for this second experiment. Three treatments were set up (Navarro-Cano et al. 2010) (i) soil without litter,

Fig. 3.4 Effects of the pine litter layer on the seedling emergence and early growth of *Diplotaxis harra* under controlled conditions

taken from the PC microenvironment (BARE); (ii) intact soil and litter forming two distinct layers as in the field taken from the BP microenvironment (PINE); and (iii) as (ii) but soil and pine litter were mixed in the laboratory (PINEMX). From the comparison of the results between treatments (ii) and (iii) it is possible to infer the different effects of a physical barrier with or without alelopathic effects (treatment ii) from purely alelopathic effects (treatment iii).

The three species do germinate significantly more on Petri dishes than in each of the treatments, something that is probably related to the presence of pathogens and allelopathic substances on natural substrates. When comparing between treatments, seedling emergence of *Diplotaxis harra* was significantly higher on BARE than on PINE or PINEMX. In respect of *Thymus zygis* there was a 50% decrease in seedling emergence on PINE in respect of BARE and more than 70% of decrease on PINMX in respect of BARE. Seedling biomass, length and number of leaves were also reduced by the presence of litter. On the contrary, *Teucrium capitatum* was weakly affected by the presence of litter, mixed or not with soil.

From this experiment it is confirmed the negative effects of pine litter on plant growth conditions of other species, at least in respect to the seedling establishment phase. As results for *Diplotaxis harra* (Fig. 3.4) and *Thymus zygis* show, this effect is not only caused by physical barrier effects, but is also an allelopathic effect. In this way, when the physical barrier effect of the litter was eliminated (PINEMX), seedling emergence was still as low as in PINE (*Diplotaxis harra*) or even lower (*Thymus zygis*).

3.3.2 Croplands

Water availability is a limiting factor for plant growth in semi-arid, rainfed conditions. The key issue of cover crop implementation is in how far the cover crop competes for water with the main crop and in how far this competition affects the economic feasibility of the cropping system. Therefore, our investigations were focused on water availability as a function of climate, soil properties, topography and cropping system.

At the European scale, climate is the most important factor that determines water availability and the potential for the application of cover crops. If the climate is drier and water availability lower, individual trees need a larger volume of soil to meet their water requirement. Hence, at sites with lower water availability, the trees need to be spaced further apart for optimal production. Climate was expected to be the most important parameter to explain the spatial variation in tree spacing at the European scale. This hypothesis was tested using a dataset of the projected canopy cover and tree density in the Mediterranean. The dataset was compiled from various sources, including field work in Southeast Spain, literature and the consultation of colleagues in Italy (contractor CNR), Spain, Portugal and Syria. The interpretation of the dataset and the implications in terms of physical thresholds for cover crops are reported in Chap. 4. The results show a significant correlation between tree density and climate (using a humidity index), this placing a limit on the extent to which cover crops can be applied for a given region.

The role of lithology on the water balance and occurrence of almond orchards was investigated at the province scale in Murcia, Southeast Spain. By courtesy of CSIC we were able to obtain a dataset on the occurrence of almonds of 1 ha resolution as well as a dataset on lithology at a resolution of 1 km^2. Lithology is directly related to the properties of marginal soils such as water retention capacity. The almond and lithology data were combined to test the hypothesis that there is a preference for almond orchards to be located on soils with a higher water availability. The results show a marked concentration of almond orchards on marls, shales and schists and an absence on limestone and dolomites.

The concentration of orchards on marls, shales and schists is probably related to the mouldability of the rock that facilitates the planting of new trees and clean-sweeping of the orchards as well as a higher water availability compared to limestones and dolomites. The second point illustrates that lithology plays an important role in terms of water availability and would therefore make a difference for the growth of cover crops. The effect of lithology on the soil water balance and the water availability for cover crops was studied in detail by quantitative modelling of the water balance on stony soils and soils on marls (see Chap. 4).

The spatial variability in water availability at the field scale was investigated by a combined field survey of almond tree vigour and patterns of overland flow. The aim was to study the importance of patterns of water availability related to topography and runoff on vegetation growth. In the literature, the trunk diameter of olive trees has been related successfully to soil properties and topography (Aragüés et al. 2004; Gálvez et al. 2004). In turn, soil properties and topographic parameters are correlated to soil moisture (Gómez-Plaza et al. 2001; Western et al. 1999). It was therefore assumed that the spatial variation in trunk diameter in young and homogeneously planted almond groves reflects the spatial variation in available water. Trunk diameter measurements were conducted in four rainfed orchards in Murcia, of which two were located on stony soils on slate and phyllites and two on marls.

On marl soil *ESV* (effective spatial variance) as measured by variograms was around 40 % and on stony soil around 24 %. Hence, the relative spatial variation in tree growth on marl soil is considerably larger. Probably this is related to a larger spatial variation in soil moisture. Nevertheless, there are also some other factors that can influence the shape of the variograms. For example, the effects on tree growth of pruning and fertilisation might not be homogeneous throughout the plantation.

A second step consisted of a study of the correlation between trunk diameter and topographic indices. A topographical survey was made for a total of five almond orchards in Murcia. The position of the almond trees, field and terrace borders were recorded with a total station and differential GPS. Subsequently, digital elevation models (DEMs) were constructed. The topographic variables considered were slope, aspect, profile curvature, tangential curvature, specific upslope area (ln$a$) and wetness index (*WI*). These parameters showed low correlations with trunk diameter, never exceeding 0.19. The low correlation between the terrain indices and trunk diameter implies that runoff patterns and storm events are of minor importance to the variation in tree growth within these orchards, probably because of the dry climate.

The hypothesis that runoff events are rare and that runoff quantities are limited is supported by rainfall measurements during the period from February 2005 to November 2006. During this period, the 10 min rainfall intensity exceeded the saturated hydraulic conductivity of the almond fields just six times. The last event was extreme with 92 mm of rainfall in 3 days, having a recurrence interval of approximately 9 years. The amount of runoff produced in the almond fields on marls remained rather limited. The results indicate that the upper limit of water availability, biological productivity and vegetation growth is determined by climate. Nevertheless, variability at finer scales (province, catchment, field) can be considerable as well, governed by lithology, soil properties and topography. This has direct consequences for the cultivation of rainfed crops and the potential for the application of cover crops: the lower the water availability, the more restricted the possibilities for the growth of cover crops.

3.3.3 Semi-Natural and Abandoned Lands

The physical characteristics and coordinates of each of 134 plots were described and all species were identified and their cover was estimated. In total 92 different species were identified of which 33 occurred in ten or more plots, which were used in the statistical analyses. The plots were chosen more or less evenly on different substrates (Quaternary colluvium, Tertiary marl, Cretaceous marl, Muschelkalk limestone, Jurassic dolomite and Keuper) and different land uses (abandoned land, semi-natural vegetation, natural forest and reforestation). With the GPS-coordinates the topographical variables (altitude, slope, position and solar radiation) were derived from a DEM for each plot (Fig. 3.5).

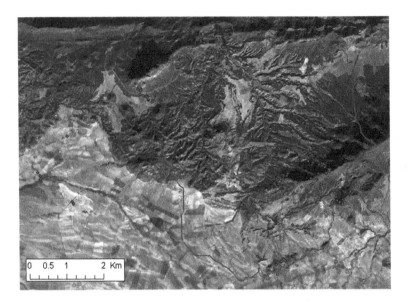

Fig. 3.5 Location of all vegetation plots in the Cárcavo basin

The species data was available in binary form (presence-absence), but also as estimated fraction of the vegetation cover. For the statistical analysis the following environmental variables were included: land use, substrate, topographical position, altitude, slope, radiation, soil depth, micro-topography and bare ground fraction. In SPSS we used the independent samples t-test (for continuous variables) and the Pearson's chi-square test (for binary variables) to test which environmental variable is significantly related to each vegetation species. Table 3.7 shows for each species the resulting significant positive or negative associations with environmental variables and in Table 3.8 the mean values of the continuous variables per species are presented.

A Principal Component Analysis of the main vegetation species and the environmental factors was conducted in which the binary variables were converted to percentages of occurrence. Four different components were extracted that explain 83 % of the variance (Table 3.9). The resulting scatter plot (Fig. 3.6) of the first two axes, which explain 67 % of the variance, shows that two groups of species can be identified. One group (*Atractylis h.*, *Diplotaxis l.*, *Artemisia h.*, *Plantago a.* and *Poa a.*) represents typical species on abandoned land, while the other group (*Quercus c.*, *Cistus c.*, *Cistus a.*, *Lithodora f.*, *Lapiedra m.* and *Juniperus o.*) corresponds to species in natural forest on steep and high positions, i.e. early against late successional species.

For the independent samples t-test the plots should be more or less randomly chosen, however due to the sampling scheme and the distribution of land uses and substrates this is not the case. Besides some of the environmental variables are highly correlated, e.g. abandoned land and Quaternary colluvium and Tertiary marl

Table 3.7 Species and significant associations with environmental variables

Species	Type	Positive association	Negative association
Stipa tenacissima	Grass	Bare ground*	Abandoned land**
			Quaternary colluvium*
			Soil depth*
Rosemarinus officinalis	Shrub	Reforestation*	Abandoned land***
		Limestone*	Quaternary colluvium***
		Top position*	Flat position***
		Micro-topography*	Radiation**
			Soil depth*
Pinus halepensis	Tree	Natural forest**	Abandoned land***
		Reforestation***	Semi-natural vegetation***
		Cretaceous marl***	Quaternary colluvium***
		Top position*	Keuper*
		Altitude**	Flat position***
		Slope**	Radiation*
		Micro-topography**	Bare ground*
Thymus vulgaris	Shrub	Cretaceous marl*	Keuper**
			Limestone*
Brachypodium retusum	Grass	Natural forest*	Keuper*
		Cretaceous marl***	Radiation***
			Bare ground*
Cistus clussii	Shrub	Jurassic dolomite**	Abandoned land*
		Muschelkalk limestone*	Tertiary marl***
		Altitude***	
		Micro-topography**	
Helianthemum sp.	Shrub	Tertiary marl*	Abandoned land*
		Jurassic dolomite**	Quaternary colluvium*
		Slope**	Keuper*
			Soil depth*
Anthyllis cytisoides	Shrub	Keuper**	
Sedum album	Herb	Natural forest*	Abandoned land*
		Jurassic dolomite**	Quaternary colluvium**
		Valley position**	Flat position*
		Slope**	
Atractylis humilis	Herb	Abandoned land**	Jurassic dolomite***
		Quaternary colluvium***	Altitude**
		Flat position*	Slope***
			Micro-topography***
Teucrium capitatum	Shrub	Reforestation***	Semi-natural vegetation**
			Muschelkalk limestone**
Helichrysum decumbens	Shrub	Keuper**	Slope*
		Bare ground*	

(continued)

Table 3.7 (continued)

Species	Type	Positive association	Negative association
Asparagus horridus	Herb	Tertiary marl***	Cretaceous marl**
		Slope position*	Top position*
		Radiation***	Altitude***
Rhamnus lycioides	Shrub	Jurassic dolomite**	Keuper*
		Slope***	Soil depth**
Quercus coccifera	Tree	Natural forest***	Abandoned land*
		Jurassic dolomite***	Tertiary marl**
		Muschelkalk limestone**	Keuper**
		Top position**	Slope position*
		Altitude***	Soil depth**
			Bare ground***
Juniperus oxycedrus	Tree	Natural forest***	Quaternary colluvium*
		Jurassic dolomite***	Tertiary marl**
		Top position***	Keuper**
		Altitude***	Slope position***
		Slope***	Radiation*
			Soil depth**
			Bare ground*
Salsola genistoides	Shrub	Tertiary marl**	Jurassic dolomite**
		Soil depth*	Altitude***
			Micro-topography*
Asphodelus fistulosus	Herb	Keuper***	Cretaceous marl**
		Top position*	Slope position*
Lapiedra martinezii	Herb	Natural forest***	Quaternary colluvium*
		Jurassic dolomite***	Soil depth**
		Altitude**	Bare ground**
		Slope**	
Pistacia lentiscus	Tree	Jurassic dolomite***	Reforestation*
		Valley position*	Tertiary marl**
		Altitude***	Soil depth*
		Slope*	Bare ground**
Juniperus phoenicea	Tree	Slope**	Radiation**
			Soil depth*
			Micro-topography*
Stipa capensis	Grass	Jurassic dolomite**	Keuper*
		Muschelkalk limestone**	Soil depth*
		Top position*	
		Altitude***	
		Micro-topography*	
Herniaria fruticosa	Shrub	Keuper***	Tertiary marl*
		Bare ground*	Jurassic dolomite*
			Slope*

(continued)

Table 3.7 (continued)

Species	Type	Positive association	Negative association
Cistus albidus	Shrub	Natural forest***	Tertiary marl*
		Cretaceous marl*	Slope position*
		Jurassic dolomite***	Radiation***
		Top position*	Bare ground***
		Altitude***	
		Slope***	
Lithodora fruticosa	Shrub	Semi-natural vegetation*	Reforestation***
		Natural forest**	Tertiary marl*
		Jurassic dolomite***	Cretaceous marl*
		Top position**	Slope position*
		Altitude***	Soil depth**
		Slope*	
Teucrium pseudochamaepitys	Shrub	Slope position*	
Plantago albicans	Herb	Abandoned land***	Reforestation*
		Quaternary colluvium**	Slope***
		Flat position**	Micro-topography***
		Soil depth*	
Diplotaxis lagascana	Herb	Keuper**	
Thymus zygis	Shrub	Quaternary colluvium*	Soil depth*
		Muschelkalk limestone***	
Helianthemum almeriense	Shrub	Semi-natural vegetation***	Reforestation*
		Keuper**	
		Muschelkalk limestone*	
Ononis tridentata	Shrub	Reforestation**	Altitude**
		Keuper***	
		Soil depth*	
		Micro-topography***	
Artemisia herba-alba	Shrub	Abandoned land*	Altitude*
			Slope*
			Micro-topography***
Poa annua	Grass	Abandoned land***	Top position*
		Quaternary colluvium**	Altitude***
		Tertiary marl*	Slope**
		Flat position***	Micro-topography**
		Soil depth***	

*$P<0.05$, **$P<0.01$, and ***$P<0.001$

Table 3.8 Mean values for the continuous environmental variables

Species	N	DEM	Slope	Radiation	Soil depth	Micro-topography	Bare ground
		m	Degrees	MJ cm^{-2} year^{-1}	1–4*	1–3*	%
Stipa t.	119	426	20.2	0.79	2.1	2.0	36.5
Rosemarinus o.	111	433	20.3	0.77	2.1	2.0	34.4
Pinus h.	83	451	22.1	0.75	2.1	2.1	31.9
Thymus v.	77	429	18.3	0.79	2.1	1.9	32.4
Brachypodium r.	65	446	20.7	0.72	2.2	2.0	30.7
Cistus c.	57	492	20.4	0.77	2.0	2.2	36.1
Helianthemum sp.	53	429	23.3	0.79	1.9	1.9	34.5
Anthyllis c.	51	437	21.2	0.82	2.0	2.0	37.0
Sedum a.	46	418	23.5	0.79	2.0	2.1	38.3
Atractylis h.	41	396	13.2	0.80	2.3	1.5	39.4
Teucrium c.	38	406	19.2	0.75	2.3	2.1	36.2
Helichrysum d.	36	434	16.3	0.74	2.2	1.9	41.5
Asparagus h.	33	358	18.0	0.87	2.1	1.8	39.1
Rhamnus l.	29	452	26.7	0.80	1.8	1.9	30.7
Quercus c.	29	552	23.2	0.73	1.8	2.1	19.3
Juniperus o.	27	532	27.5	0.69	1.7	2.0	27.6
Salsola g.	25	350	19.3	0.75	2.6	1.6	40.0
Asphodelus f.	25	407	18.1	0.74	2.1	2.0	41.6
Lapiedra m.	25	489	25.1	0.73	1.7	2.3	24.0
Pistacia l.	24	531	24.0	0.84	1.8	2.3	25.2
Juniperus p.	21	408	27.2	0.67	1.8	1.6	33.1
Stipa c.	20	539	23.5	0.79	1.8	2.4	28.0
Herniaria f.	16	410	14.0	0.77	2.4	2.2	46.6
Cistus a.	15	529	29.7	0.56	2.2	1.8	14.7
Lithodora f.	15	538	25.5	0.78	1.4	1.8	37.3
Teucrium p.	15	399	20.7	0.87	2.1	1.7	26.3
Plantago a.	14	398	9.2	0.78	2.8	1.2	37.5
Diplotaxis l.	13	396	14.6	0.76	2.3	2.1	42.3
Thymus z.	13	450	20.0	0.79	1.7	1.8	25.0
Helianthemum a.	12	425	15.3	0.78	2.1	1.6	43.3
Ononis t.	11	335	19.8	0.68	2.7	2.6	41.8
Artemisia h.	11	388	14.4	0.76	2.5	1.1	38.2
Poa a.	10	334	9.3	0.82	3.5	1.3	41.0

*Soil depth (1 = <10 cm, 2 = 10–25 cm, 3 = 25–50 cm and 4 => 50 cm) and micro-topography (1 = low, 2 = medium, 3 = high)

and reforestation. Therefore, a careful interpretation of the analysis is required, e.g. a very significant negative relation between a certain species and a certain environmental variable indicates a low probability of occurrence on such a location, however, it does not mean that such a species cannot grow in that environment. For example, *Pinus halepensis* had a very significant (P < 0.001) negative relation with

Table 3.9 Component factors from PCA

Environmental variable	Axis 1	Axis 2	Axis 3	Axis 4
Altitude	−0.497	0.733	−0.236	0.054
Slope	−0.103	0.797	−0.457	−0.070
Radiation	0.029	−0.340	0.141	0.888
Soil depth	0.285	−0.679	0.405	−0.317
Micro-topography	0.161	0.658	−0.021	−0.009
Bare ground	0.257	−0.608	0.465	−0.003
Abandoned land	0.423	0.571	0.632	−0.086
Semi-natural vegetation	0.946	0.060	0.009	−0.048
Natural forest	0.793	−0.270	−0.440	−0.077
Reforestation	0.958	0.066	−0.080	0.134
Quaternary colluvium	0.674	0.430	0.525	−0.089
Tertiary marl	0.839	0.217	0.060	0.026
Keuper	0.697	0.207	0.000	0.351
Cretaceous marl	0.908	0.013	−0.058	0.049
Jurassic dolomite	0.784	−0.358	−0.362	−0.225
Muschelkalk limestone	0.823	−0.040	−0.262	0.118
Flat position	0.574	0.537	0.445	−0.039
Valley position	0.852	−0.128	−0.191	−0.213
Slope position	0.961	0.155	0.058	0.086
Top position	0.941	−0.115	−0.267	0.054

abandoned land, but this can be explained by the slow secondary succession and the absence of reforestation on abandoned land. Especially on abandoned land and in reforested areas, the human influence has highly disturbed the natural occurrence of vegetation species.

3.3.4 Channels

Recognising that the variation in mapped vegetation reflects a range of conditions, further site investigations were undertaken to determine the conditions for growth and survival of plants within each assemblage. Site selection was based on the maps of vegetation assemblages, such that there was sufficient number of sites within each of the assemblage categories identified. A site is an area of vegetation within a reach and cross-section. There may be more than one zone of vegetation within a cross-section and each is a site. In total, 39 sites were selected for detailed study along Cárcavo (c.7 km in length) and 36 along Torrealvilla (c.6 km in length). At each site a number of variables were measured. The choice of variables measured was based on those variables identified in the literature as possibly exerting a controlling influence on the composition and distribution of riparian vegetation (Baker 1989; Birkeland 1996; Craig and Malanson 1993; Harris 1987; Hupp and Osterkamp

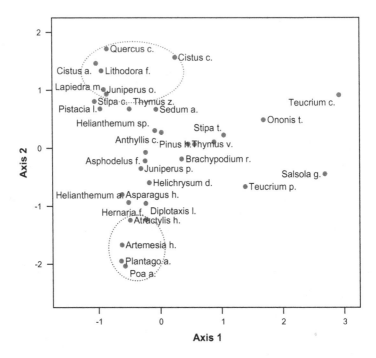

Fig. 3.6 Scatter plot of the principal component axes

1985; Malard et al. 2002; McBride and Strahan 1984; Osterkamp and Hupp 1984; Tabacchi et al. 1996; Zimmerman 1969; Zimmerman and Thom 1982). Variables measured can be grouped into four main categories: water availability, substrate, geomorphology and hydraulics.

Availability of water has been identified as a key factor controlling the distribution of vegetation in semi-arid channels (Tabacchi et al. 1996; Zimmerman 1969). Studies of vegetation patterns along ephemeral channels in SE Spain by Tabacchi et al. (1996) indicated that longitudinal change in water supply was probably the most important factor in structuring riparian communities. Zimmerman and Thom (1982) showed how vegetation patterns reflected changes in surficial geology and the influence that this has on surface water/groundwater flows, with marked increases in vegetation diversity associated with sections of the channel where there was a perched aquifer with baseflow on a bedrock aquiclude. Hence, the thickness of sediments and/or depth to bedrock is also considered to be important in influencing water availability along the channel.

A means of measuring both the depth to bedrock and depth to water was therefore considered important. Where it was possible, a hand auger was used to core through the bed to bedrock. The thickness of the unconsolidated sediments, depth to bedrock and water table was then determined with a measured tape. However, it was not always possible to core to bedrock, such as in circumstances where there were coarse gravels, or the substrate was loose and could not be brought to the surface.

For sites positioned on gravel bars and where bedrock was exposed in the thalweg, the vertical distance of the site above bedrock was measured and used as a value of depth to bedrock and water. Where bedrock was not exposed, the vertical distance of the site above the thalweg or depth of sediments augered was used as a minimum estimate of water table depth. Where the site was situated in close proximity to a check dam, it was possible to estimate the depth to bedrock by measuring the height of unconsolidated sediments behind the check dam.

Variation in *sediment characteristics* is considered to be one of the most important factors influencing plant establishment and survival (McBride and Strahan 1984; Osterkamp and Hupp 1984). The texture of channel sediments at the surface were described in the field and classified into 16 simple and mixed classes Where gravels and boulders were present, the mean size and range of clasts were measured. Where finer channel sediments were present, sediment samples were collected and analysed in the laboratory for particle size, carbon content and aggregate stability. In the process of augering to determine the depth to bedrock and water table, the stratigraphy of the channel bed was also described, documenting the variability in thickness and calibre of sediment layers.

Different *landforms* within the channel are characterised by a particular flow and sediment regime that has led to their development, and the regime associated with each landform may support a particular assemblage of plants (Bendix and Hupp 2000). Therefore, the overall morphology and positioning of the vegetation assemblage within the channel is of importance in influencing conditions. The overall morphology of the site was classified as: 1. flat bed, 2. flat bed with vegetation wakes, 3. thalweg, 4. mid-channel bar, 5. side bar, 6, point bar, 7. floodplain and 8. bedrock. The lithology of the valley walls varies markedly along the course of the channels and may influence the characteristics of the sediment supplied to the channel. The lithology of the base of the valley walls was distinguished. The dimensions of the channel (width, depth) and valley (width) were measured in the field and channel width/valley width and channel width/depth ratios then determined. Inner and outer channels were defined in some cases and tests were made with mean and maximum depths and of widths at various heights.

One of the basic premises of the RECONDES project is that vegetation is influenced and in turn has an influence on the *connectivity* of flow and sediment transfers within the drainage basin. The potential connectivity of each site has been assessed through mapping of morphology, vegetation and connectivity along the channel using the methods described by Hooke (2003). This method of mapping is based on the interpretation of various morphological and sedimentological evidence. In this, the morphology of the channel is mapped, detailing changes in the dimensions of the channel bed and floodplain, bedrock exposures and the position of check dam/groyne structures and road crossings. Particular attention is given to mapping the sources (gully/tributary, banks/valley walls) and storages of sediment (bar forms, accumulations behind check dams) along the channel. Using the combined map layers (channel morphology, sediment sources and storages) sediment source zones are identified and the channel is divided into areas of erosion, erosion and storage and net storage. The potential connectivity of sites along the channel network was then assessed and classified. Reaches that are classified as 'area of net erosion' are highly competent, material is likely to be flushed through the reach at high flows and there is minimal potential for sediments to be deposited within the reach. Within reaches classified as 'erosion and storage' there is considerable exchange of sediments along the reach. For example, sediments may be deposited and stored through the growth of one bar but a similar volume of sediments is then eroded from the next bar downstream. Within sections of the channel classified as 'areas of net storage' there is evidence of continued build up of sediments. Reaches with 'stable vegetation' typically extend upstream of check dams and represent reaches where the channel has been stabilised and vegetation now covers the channel floor.

The *hydraulics* of floods play an important role in influencing the distribution of vegetation. In-channel vegetation is affected by flood processes such as scour and deposition during flows and the characteristics of the landforms that are formed during floods (Bendix and Hupp 2000). Feedback relationships between vegetation, channel morphology and hydraulics exist, where the morphology of the channel and vegetation assemblages both influence and are influenced by flow hydraulics (Hooke and Mant 2000). Different landforms within the channel are characterised by a particular flow regime that has led to their development, and the flow regime associated with each landform may support a particular assemblage of plants (Bendix and Hupp 2000). Detailed cross-sections were surveyed for each of the sites using a Differential GPS (±2 cm) and recording changes in substrate and vegetation (species, height, condition) along the cross-sections. A longitudinal profile was also surveyed from which longitudinal distance, elevation and gradient at each site could be derived. Cross-sectional profile and gradients are used in conjunction with detailed surveys of vegetation to estimate hydraulics at sites for a selection of flows. The lack of multiple cross-sections at each site precluded use of software such as HEC-RAS. WinXSPRO (Hardy et al. 2005), was chosen for the calculation of hydraulics because of its ability to compute flow hydraulics for single cross-sections and the ability to divide a cross-section into a number of subsections on the basis of morphology or vegetation in which hydraulics could be calculated. The program has been recently used to compute hydraulics of flow in a study by Lite et al. (2005) examining the influence of flood disturbance and water stress on riparian vegetation along the San Pedro River in Arizona, a similar type of situation to that in the Spanish channels.

Cross-sections were divided into a number of subsections based on differences in vegetation and topography and Manning's n values were estimated for each subsection, according to the methods of Arcement and Schneider (1989). The program calculates the velocity and discharge of flows within each of the subsections for the increments of flow stipulated. Hydraulics of flows were calculated at selected sites for low, medium and high discharge events along Cárcavo (2, 10 and 40 $m^3 s^{-1}$) and Torrealvilla (10, 100 and 200 $m^3 s^{-1}$). These discharges were selected as having ecological and geomorphological significance. The low flow fills the thalweg/inner channel, the medium flood most closely resembles that of bankfull and the high flow is at least twice the magnitude of the medium flow. The bankfull stage is somewhat difficult to prescribe in some places and very sparse information is available on the frequency and magnitude of flows. Monitoring of flows within the Guadalentín Basin by one of the authors over the past 10 years also informed selection of these flows.

This method of calculating flow hydraulics provides the basis for evaluating the effect that different vegetation assemblages have on flow hydraulics, the conditions in which particular vegetation assemblages exist and flows they can withstand. The results can then be compared with surveyed effects of flows on vegetation to assess the thresholds of forces for plant damage and destruction. For this, we draw upon some earlier monitoring work that was completed in the region, in particular, the impact that floods in September 1997 and October 2003 within the Guadalentín Basin and November 2006 in Cárcavo Basin have had on vegetation and channel morphology.

Statistical analysis of the environmental data was carried out with the aim of identifying the conditions for the establishment and survival of different species and assemblages and thresholds. Principal Component Analysis (PCA) was used to eliminate redundant variables and determine which variables are more influential than others. PCA was carried out in a number of steps. In the first instance variables were grouped together according to whether they were considered to influence water availability, substrate, morphology and hydraulics. PCA was then carried out on each of these groups (Table 3.10). PCA was carried out on Cárcavo and Torrealvilla datasets separately, and also on the combined dataset. The analysis of the combined datasets gives considerably higher factor loadings so these results are presented.

The results of Principal Component Analysis (PCA) for water availability, substrate and morphology are shown in Fig. 3.7. For each PCA the set of variables has been reduced to two components, Component 1 and 2. Component scores are plotted on x and y axes according to the dominant plants. This is informative in revealing whether sites with the same vegetation plot together which would indicate similarity in required conditions for establishment and/or there are distinct differences in conditions associated with sites of contrasting vegetation types. The extent to which similarly vegetated sites are distributed in space is also informative in revealing the range of tolerances associated with each vegetation type. The amount of influence a variable has on each component is indicated by the value of its factor loading, listed in the Table 3.11 for water availability, substrate and morphology. The percentage variance explained by each PCA is also indicated in these tables.

For hydraulics, it is necessary to complete the PCA on the Cárcavo and Torrealvilla datasets separately as the low, medium and high flows computed are not the same for both channels. The Torrealvilla drains a much larger catchment area than Cárcavo and is able to generate significantly larger floods. Fig. 3.8 shows the output of PCA on the high flow computed for Cárcavo and Torrealvilla. There is a fairly clear grouping of *T. canariensis* and *N. oleander* together, this attributed in part to their comparable roughness values. Sites with no vegetation and grasses have comparably high flow velocities, these being located in high velocity areas, closer to the thalweg where flows reach greater depths.

Table 3.10 List of independent variables and groupings chosen for principal component analysis

Code	Description	Water availability	Substrate	Morphology	Hydraulics
Relev_site	Height above thalweg (m)	X	X	X	X
SDelev_site	Standard deviation of site (m)	X		X	X
Slope	Slope at site	X	X	X	X
Cd_up	Distance to checkdam upsteam (m)	X	X	X	X
CD_do	Distance to checkdam downstream (m)	X	X	X	X
BC_width/ V_width	Bankfull width/Valley width ratio			X	
IC_width/ IC_Adepth	Inner channel width/ Inner channel depth ratio			X	
BC_width/ BC_A depth	Bankfull width/depth ratio		X	X	
VWB_lith	Lithology of valley wall		X		
Chann_morph	Morphology of channel, position of site with respect to channel			X	
Substrate	Substrate – field textural classification (1–16)	X	X		
Re_Substrate	Simplifed substrate – fines dominant, gravels dominant, bedrock dominant	X	X		
D_bedrock	Depth to bedrock (m)	X			
D_water	Depth to water (m)	X			
Carbon	% Carbon from surficial sediments		X		
text_gravel	% Gravel		X		
Text_sand	% Sand		X		
Text_mud	% Mud		X		
So_d	Mean depth (section)				X
So_v	Velocity (section)				X
So_totp	Total Power (section)				X
So_unitp	Unit Stream Power (section)				X
So_br	Blockage ratio				X
So_ir	Inundation ratio – height of water/height of plants				X
So_ss	Shear stress (section)				X
To_q	Discharge across cross-section for flood stage (total)				X

(continued)

Table 3.10 (continued)

Code	Description	Water availability	Substrate	Morphology	Hydraulics
To_xsa	Total cross-sectional area (total)			X	X
To_v	Velocity (total)				X
To_totp	Total Power (total)				X
To_unitp	Unit Stream Power (total)				X
To_ss	Shear Stress (total)				X
Status	Index of connectivity		X		

Review of the PCA for water availability, substrate and morphology indicated that substrate (% sand, % gravel), index of erosion/connectivity, depth to bedrock, and height above thalweg were most important in explaining the variance of conditions (as indicated by their high factor loadings). A final PCA was conducted on these variables. This PCA showed very clear separations between *T. canariensis* and *N. oleander* dominated sites and the different density of grasses (Fig. 3.9). Reeds appear across a range of conditions, but a larger proportion of the sites with Reed dominated sites appear to coincide with areas of net storage, fine substrates, and greater depths to bedrock (these conditions are satisfied upstream of check dams).

In summary, Principal Component Analysis of the environmental dataset indicates that substrate is particularly important, but linked with this is also the index of potential connectivity/overall status of the reach. *T. canariensis* are rooted in fine substrates, sites where they are located are classified as areas of net storage, this coinciding with areas upstream of check dams where sedimentation of fines occurs. In contrast *N. oleander* are rooted in gravel substrates, and within the channel network these are mapped as areas of erosion, sediment transport and storage, tending to be part of a larger bar structure or former channel surface. Grasses do occupy a range of substrates. However, for a high density of grass cover to establish, fine sediments are important and these sites fall within reaches that are classified as areas of net storage. The second factor which is considered important is that of water availability, though the distinctions between the different vegetation types are not clear. This is probably in part, a problem of the difficulty associated with measuring this factor in the field. It is dynamic in nature, fluctuating considerably throughout the year in response to seasonal rains and flows. It is in the early stages of plant growth when water availability is most important but mapping and selection of sites is based on the distribution of mature vegetation assemblages. The variables we have measured and included in the PCA of water availability are an index of the potential for water to be available at a site. What is revealing from this PCA is the wide range in which *T. canariensis* dominated sites are found, extending into areas where depth to bedrock and water is greatest. This is supported by what is known from existing literature that *T. canariensis* has a

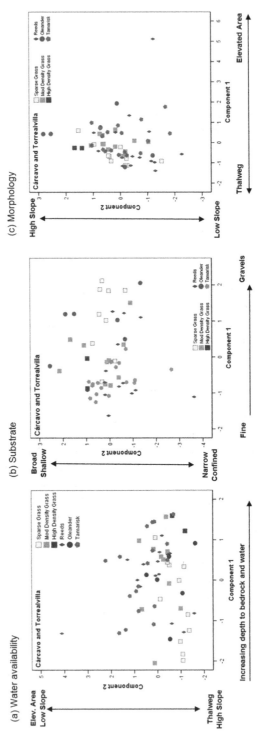

Fig. 3.7 Principal component analysis plots for (**a**) Water availability, (**b**) Substrate and (**c**) Morphology. Details of variables included in each PCA and their factor loadings are outlined below each scatterplot of components

Table 3.11 Table results of principal component analysis for (a) water availability, (b) substrate and (c) morphology

(a) Water availability

Comp.[a]	% V.E.[b]	Variables	F.L.[c]
1	37%	Depth to water	0.83
		Depth to bedrock	0.79
		Dist. to checkdam downstream	−0.69
		Simplified substrate	−0.63
2	26%	Height above thalweg	0.78
		Slope	−0.59
		Simplified substrate	−0.53
		Depth to bedrock	0.47

KMO Statistic=0.61, Total Variance Explained=63%

(b) Substrate

Comp.[a]	% V.E.[b]	Variables	F.L.[c]
1	36%	Substrate	0.87
		Status	−0.8
		% Gravel	0.78
		% Sand	−0.43
		Slope	0.43
2	17%	Bankfull width/ average depth	0.76
		Height above thalweg	−0.59
		Slope	0.48

KMO Statistic=0.71, Total Variance Explained=53%

(c) Morphology

Comp.[a]	% V.E.[b]	Variables	F.L.[c]
1	35%	S.D. of elevation	0.92
		Height above thalweg	0.92
		Bankfull width/ valley width	−0.14
		Bankfull width/ average depth	−0.11
2	27%	Slope	0.8
		Bankfull width/ valley width	0.75
		Bankfull width/ depth	0.3
		Height above thalweg	−0.17
		S.D. of elevation	−016

KMO Statistic=0.56, Total Variance Explained=62%

[a]Component, [b]% Variance Explained, [c]Factor Loading

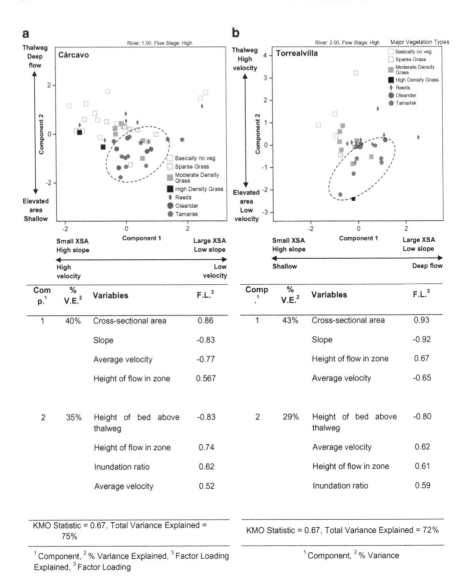

Fig. 3.8 Principal Component Analysis plots of hydraulic variables for (**a**) Cárcavo sites at high flow (40 m³ s⁻¹). (**b**) Torrealvilla sites at high flow (200 m³ s⁻¹). Details of variables included in each PCA and their factor loadings are outlined below each scatterplot of components

high tolerance to water stress (Levine and Stromberg 2001). The conditions required for reeds do not emerge clearly, and it is considered that perhaps the localised conditions where ponding occurs are not highlighted in the environmental variables measured.

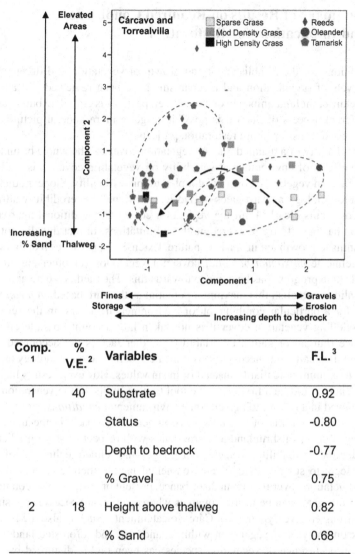

Comp.[1]	% V.E.[2]	Variables	F.L.[3]
1	40	Substrate	0.92
		Status	-0.80
		Depth to bedrock	-0.77
		% Gravel	0.75
2	18	Height above thalweg	0.82
		% Sand	0.68

KMO Statistic = 0.74, Total Variance Explained = 58%

[1] Component, [2] % Variance Explained, [3] Factor Loading

Fig. 3.9 Principal component analysis of high scoring variables showing segregation of vegetation. *Arrow* shows trend of increasing density of grasses with greater proportion of fine sediments

3.4 Summary of Results on Required Conditions and Implications for Restoration

The conditions for the establishment and growth of vegetation at different spatial scales, levels of organization and mechanisms have been reviewed and analysed across the different land units. No one of the perspectives is complete but by assembling different aspects of them it is possible to get a more general picture of the necessary conditions for future restoration projects.

The availability of a detailed map of vegetation cover for the whole basin lets us study the effects of environmental variability on vegetation cover. It is confirmed that variations in vegetation cover are coupled to this variability. Slope gradient and unfavorable lithology as surrogates of hydric, nutritional and erodibility attributes of the ecosystems clearly limit the vegetation cover. The traditional approach of restoration has been to try to overcome these limitations. In fact this did not work well as limits to growth are intrinsic to nature. Extensive approaches like this should lead to delineate *maximum potential* cover for vegetation (or other attribute like LAI) and some progress has been made towards this. The methods here applied are not yet valid to establish this maximum potential as they are based on *mean* values of cover for a particular combination of environmental factors, in the sense that when modelling vegetation cover it is not taken into account possible effects of disturbance (human or natural), variability in other factors not accounted in our survey, etc. Basically, it is necessary to delineate a statistical methodology aimed to estimate maximum potentiality instead of mean values. However, results like those presented here could be a first operative tool to plan limits on the vegetation cover to be achieved in a restoration *given* some environmental *conditions*.

The potential pitfalls of a purely environmental approach is the unexpected behaviour of cover in shrubland and grasslands with respect to slope gradient and solar radiation. Availability of new better images and refined techniques of image analysis seem to suggest that: (i) environmental factors themselves could fail to explain vegetation cover if: human disturbance, present or past, is not accounted for; (ii) used surrogates can be useless in areas where dominant processes are singular like the strong erosive dynamics of Cárcavo catchments. Such is also the finding of the work on analysis of vegetation within abandoned and reforested lands, where the natural occurrence of vegetation species has been highly disturbed by human influences.

On these grounds we advocate the routine inspection of conditions for growth of areas to be restored using high-resolution images, using simple surrogates to model vegetation attributes and then inspecting what and why they are departing from previous assumptions. It will allow identification at the ecosystem and landscape level of where the conditions are adequate and above all, what limit the conditions impose on the possible development of vegetation.

The study of the relationship between conditions and planted *Pinus halepensis* and growth pointed out clearly how environment limits the possibilities of restoration of vegetation. Homogeneous treatment of a highly variable environment led to futile efforts as in unfavorable situations (high slope gradient, low infiltration and high runoff) mortality is very high, limiting establishment. On the other hand, those trees that survive have their growth clearly controlled by hydric stress and soil status as indicated by the significant effect of radiation and lithology. This large experiment provided by former management clearly indicates that conditions should be taken into account in any future planning. There are several ways of practical exploitation of this approach. Relationship between slope gradient and mortality can be used to establish limits to maximum gradient to be planted. Another similar approach is to estimate final densities, after differential mortality and to establish different densities of plantation according to slope gradient. Also, it can be established what combinations of environmental factors produce so slow growth that they are not considered viable for the objectives of restoration. These combinations of factors would be located in the field through GIS and will not be planted with the objective species. Comparing with whole scale basin analysis it could be apparent that a clear limit to vegetation development exists under these conditions, and then only very point-focused restoration would be carried out, for instance to short-circuit connectivity of the systems.

At the micro-scale physical and chemical attributes of litter layer have been shown to hamper soil development and germination dynamics. The litter layer of dominant species may be working as a physical barrier preventing the seedling establishment and the effective rain infiltration, but also can have significant allelopathic effects on the seed germination and early growth of plants. On the other hand, the planted vegetation had a negligible effect on the soil quality in the medium term due to the combination of poor litter decomposition and adverse climatic conditions. Also, increased soil organic carbon concentration of reforested areas was counteracted by higher rates of soil organic carbon erosion in well connected slope-channel areas (Boix-Fayos et al. in press).

Altogether, *a priori* approaches to semi-arid vegetation restoration that are based on simplistic assumptions that large scale afforestation will trigger secondary succession are risky if their effect is not previously tested; they are highly questionable and are doomed to failure. More finely targeted actions may be necessary to speed up soil quality improvement and secondary succession, for example choosing species with better litter quality, or managing the litter of pines by selectively removing it to facilitate the seedling establishment of subdominant species and spreading it on rills to increase rugosity and decrease connectivity of water and sediment flows. Many unexpected results may arise from the intrinsic complexity of the ecological relationships, which need to be addressed in detail.

Water availability is the main factor limiting the growth of cover crops in croplands. Soil type is also significant in influencing water retention capacity. Variations in soil type and topography at the field scale influence water availability for cover crops and orchards. The application of cover crops may be beneficial in decreasing erosion; however, their application needs to be adapted to the management and growing season of the main crop. The main concern is over the potential for competition for water between the cover crop and the main crop leading to subsequent reductions in yield.

In Semi-natural and Abandoned lands the vegetation on 134 plots were identified covering a range of substrates and types of land use, and a range of environmental variables were measured in the field and from a digital terrain model. Principal component analysis of main vegetation species and environmental variables revealed clear separation of species found in abandoned lands from those found in natural forest areas, difference between early and late succession species. Interpretation of the analyses requires careful consideration as a strong negative relation between a particular species and environmental variable does not necessarily mean that species will not grow in that environment. For example, *Pinus halepensis* has a very significant (P<0.001) negative association with abandoned land, but this can be explained by the slow secondary succession and the absence of reforestation on abandoned land.

Principal Component Analysis techniques has resolved the main factors that are important in influencing plant establishment, and the requirements for different types of vegetation along river channels. Substrate has been shown to be particularly important, but also linked to this is the degree of connectivity/overall sediment status of reach. Fines are particularly important for establishment of the perennial tussock grass *Lygeum spartum* and *Tamarix canariensis*, whilst on coarse substrates *Dittrichia viscosa* and *Nerium oleander* are more likely to establish. Water availability is also considered important. The distribution of check dams has a strong influence on conditions for different plants to establish in the channel.

In summary, a top-down, landscape approach in studying conditions for vegetation establishment and growth could be interesting in order to produce successively improved approaches to the restoration of an area. This successively improved approach goes from ecosystem attributes at catchment scale looking for deviations of a purely resource-based ecosystem functioning that could indicate operation of disturbances to physico-chemical conditions for dominant species to subtle biological controls. To conclude, it is important to have in mind that conditions themselves impose limits to vegetation development that simply cannot be overcome by any 'restoration'. Setting these limits is a priority task.

Chapter 4
Effectiveness of Plants and Vegetation in Erosion Control and Restoration

Peter Sandercock, Janet Hooke, Sarah De Baets, Jean Poesen,
André Meerkerk, Bas van Wesemael, and L.H. Cammeraat

Abstract In this chapter the approaches and methods used to measure plant effectiveness in reducing runoff and erosion are explained and results presented for each of the major land units, hillslopes and channels. Evaluations of the properties of plants required are made to inform plant selection for different sites. For use of cover crops in orchards it is important to assess whether the cover crops would have an effect on orchard tree productivity, whilst also reducing soil erosion. A climatic threshold for their use was identified. Soil moisture measurements from different treatment areas and water balance and runoff modelling exercises showed where use of such crops could be beneficial. Extent of vegetation growth on abandoned lands was shown to have a marked effect on runoff, water repellency and soil crusts. Various root parameters were measured on a range of plants and their relation to soil detachment calculated. Differences in root architecture and in orientation of rows of plants were tested. Plant stem density, stem bending and trapping efficiency effects

P. Sandercock (✉)
Jacobs, 80A Mitchell St, PO Box 952, Bendigo, VIC, Australia
e-mail: Peter.Sandercock@jacobs.com

J. Hooke
Department of Geography and Planning, School of Environmental Sciences,
University of Liverpool, Roxby Building, L69 7ZT Liverpool, UK

S. De Baets
College of Life and Environmental Sciences, Department of Geography,
University of Exeter, Rennes Drive, EX4 4RJ Exeter, UK

J. Poesen
Division of Geography and Tourism, Department of Earth and Environmental Sciences,
KU Leuven, Celestijnenlaan 200E, 3001 Heverlee, Belgium

A. Meerkerk • B. van Wesemael
Earth and Life Institute, Université catholique de Louvain (UCL),
Place Louis Pasteur 3, 1348 Louvain-la-Neuve, Belgium

L.H. Cammeraat
Instituut voor Biodiversiteit en Ecosysteem Dynamica (IBED)
Earth Surface Science, Universiteit van Amsterdam, Science Park 904,
1098 XH Amsterdam, The Netherlands

© Springer International Publishing Switzerland 2017
J. Hooke, P. Sandercock, *Combating Desertification and Land Degradation*,
SpringerBriefs in Environmental Science, DOI 10.1007/978-3-319-44451-2_4

were also assessed experimentally and plant species growing in the Mediterranean study area were grouped according to their erosion control potential. The effects of vegetation and various plant species on roughness, flow hydraulics and sediment trapping in channels were assessed by field measurements and modelling and their resilience to high flow evaluated from observed flood impacts.

Keywords Erosion control effectiveness • Cover crops • Crop water balance • Plant root properties • Vegetated channel roughness • Flood impacts on vegetation

4.1   Introduction

Understanding the effect that plants and vegetation have on processes is critical to assessing their effectiveness in prevention of erosion and selection for use in the stabilisation of degraded lands. In this chapter the approaches and methods used to measure plant effectiveness in reducing runoff and erosion are explained. Results of each are exemplified. Evaluations of the properties of plants required are made to inform plant selection for different sites. These aspects are discussed and exemplified for each of the major land units then hillslopes and channels.

4.2   Land Units

4.2.1   Cover Crops

The main issue in relation to assessment of effectiveness of cover crops is not only its ability to reduce erosion but also to determine whether it affects productivity of the main crop, through reduction of water availability. The effectiveness of cover crops can be assessed at the field scale using a number of different approaches. Three approaches were taken within the project: (1) calculation of climatic thresholds, (2) soil moisture measurements from different treatment areas (i.e. areas with and without cover crops), and (3) water balance and runoff modelling exercises to investigate the availability of water under different climate and soil conditions.

4.2.1.1   Identification of a Climatic Threshold

Whereas climate determines the upper limit of available water, an attempt was undertaken to identify climatic thresholds for cover crops in olive orchards. This study was based on the relation between tree density, projected canopy cover and climate (Meerkerk et al. 2008). As a climatic indicator the ratio between annual precipitation and the reference evapotranspiration was used. In the literature this ratio is referred to as the aridity index (Lioubimtseva et al. 2005; UNEP 1997) or humidity index (Yin et al. 2005). The term humidity index (*HI*) is used here because this index increases with increasing humidity. It can be noted that the semi-arid climatic zone is defined by $0.2 <= HI < 0.5$ (UNEP 1997). During the project a considerable amount of data on olive orchards was gathered, both from fieldwork in Southeast Spain and from colleagues and literature sources throughout the Mediterranean. For each observation, *HI* was calculated using data from the A-Team climate database (Mitchell et al. 2002). Evapotranspiration was estimated with the Hargreaves equation, using monthly input data (for details see Allen et al. 1998).

The results produced a significant correlation between tree density or canopy cover and *HI*, which is nevertheless rather weak (Meerkerk et al. 2008). It appears that for locations with $HI > 0.6$, the growth of the olive trees is no longer limited by water availability and there is a surplus of water that could be used to grow cover crops. Below the threshold, the use of cover crops must be restricted in space and time in order to avoid competition for water. This threshold coincides with the crop factor, K_c. Probably, for other perennial crops the K_c can be used as a climatic threshold as well.

4.2.1.2 Experimental Field Study

Soil moisture measurements were set up in an olive orchard west of Sevilla (UTM 29S 745565E 4136755 N) (in collaboration with colleagues from CSIC Córdoba) (Fig. 4.1). The experiment consists of the comparison of two treatments: (1) no-turning tillage (3–5 times per year) and (2) a grass cover in the traffic lane that is killed in spring with herbicides. The soil moisture is measured at depths of 7, 15, 21 and 45 cm using theta probes. Two sensors were installed per treatment at 45 cm. Only preliminary results were obtained within the period of the RECONDES project and are not discussed here.

4.2.1.3 Water Balance and Runoff Modelling

The aim of the water balance modelling exercise was to compare the effect of different climatic and soil conditions on water availability in rainfed cropping systems. It serves to improve our knowledge on the water availability for cover crops and competition with the main crop (Meerkerk et al. 2008).

For the almond orchards studied in Murcia, a two-layer approach is used in order to distinguish evaporation losses from the plough layer and the water in the underlying soil, which is available for transpiration by the trees. This conceptual model builds on two assumptions. The first assumption is that no drainage occurs below the rooting zone of the almond trees. This is realistic under semi-arid conditions, since the amount of rain from individual rainfall events is small. For example, in the Cárcavo target area, a 35-year record of weather data reveals that the rainfall exceeds

Fig. 4.1 The experimental
site west of Sevilla. Upper:
no-turning tillage. Lower:
cover crop treatment
(Meerkerk et al. 2008)

30 mm only during 3 % of the rainy days. Furthermore, the roots of full grown trees may extend up to 14 m from the trunk and a few meters deep if the soil is deep enough (Micke 1996). This means that although the trees are typically spaced about 7 m apart, their roots can potentially extract water from the complete orchard surface, limiting drainage. The second assumption of the model is that no almond roots occur in the 15 cm deep plough layer; the soil is kept bare during the major part of the year by frequent tillage.

In order to verify the validity of these assumptions, a survey was done of the presence of roots in the upper 25 cm of the soil in an almond orchard in Cárcavo. In total 42 root observations were done at distances of 0.5–4 m from the trunks, examining vertical 10 by 10 cm squares of soil at a depth of 15 and 25 cm. In 11 pits (25 %) no roots were encountered at all, and in just 2 pits (5 %) there were more than ten roots per 100 cm² window. Roots with a diameter beyond 1 mm were observed in seven pits (17 %), of which six pits had a single root of that size and one pit had two. Sometimes the roots could be related to weed remnants, supporting the conclusion that the presence of tree roots in the plough layer is quite restricted.

Modelling was used to test the hypothesis that the evaporation loss from the plough layer (E_p) can be used without competition with the trees by a shallow rooting cover crop (Meerkerk et al. 2008). In other words, the deep rooting almond crop and shallow rooting cover crop derive their water from different parts of the soil. The water balance of the plough layer can be written as:

$$E_p = P - D - \Delta S - Q_{off} + Q_{on} \tag{4.1}$$

where E_p is evaporation loss from the plough layer (mm), P is precipitation (mm), D is drainage below the plough layer(mm), ΔS is change in soil moisture storage for a given time interval (mm), Q_{off} is runoff (mm), Q_{on} is run-on (mm).

   Two soil types were selected for a simulation exercise, consisting of a silt loam soil on marl and a stony soil on slate and phyllites. The contrasting properties of these soils allow us to obtain a broad picture of the water balance and evaporation loss from the plough layer in Murcia. The BUDGET model (Raes et al. 2006), is able to simulate the water balance of both soil types quite well. BUDGET was calibrated by means of a lysimeter experiment which included two soil columns per soil type with a diameter of 30 cm. For the calibration, three parameters of the model needed to be adjusted for each soil column: depth of evaporation, field capacity or wilting point and tau, which determines the rate of drainage within the soil profile. The simulation errors of the calibrated model were smaller than 11 mm for all components of the water balance and the root mean square error in soil moisture was below 4.6 %. Runoff was estimated based on the initial abstraction before runoff and the saturated hydraulic conductivity of the soil. An empirical relation was used that estimates the maximum 10-min rainfall intensity from hourly rainfall records ($R^2 = 0.66$; $n = 118$), as well as the intensity of the rain that falls during the remaining 50 min.

   Model runs were executed with hourly weather data measured by UvA at a field site 16 km northwest of Lorca (Alqueria station). Next to a year with average rainfall (287 mm), a dry (169 mm) and wet (429 mm) year were selected, having a return period of approximately 7 years. This range in annual rainfall should be considered in the design and management of the rainfed orchards and cover crops, in order to avoid an unsustainable tree mortality rate in dry years.

   A part of the model input data was derived from field measurements that were carried out in Cárcavo in January and September 2005. The dry bulk density of the top soil ranged from 1000 to 1200 kg/m³. The hydraulic conductivity of the top soil (K_{sat}) was determined by rainfall simulation on freshly ploughed soils and soils with a structural crust. The freshly ploughed soils developed a structural crust after just one rainfall event. The observed K_{sat} values ranged from 18 to 31 mm/h. The average for non-crusted soil was 29 mm/h and for crusted soil 23 mm/h. Another input was the soil albedo, which was measured during fieldwork in January 2006.

   The results of the simulation study show that the evaporation loss is 52–112 mm higher on the marl soil for the 'wet' and 'average' year compared to the stony soil. This difference is directly related to the deeper infiltration of the rain on the stony soils. In the 'average' and 'dry' year, there is no drainage below 33 cm on the marl, due to the small size of the rains (<=41 mm) and low antecedent soil moisture. Another difference is the occurrence of runoff on the marl soils, which increases the

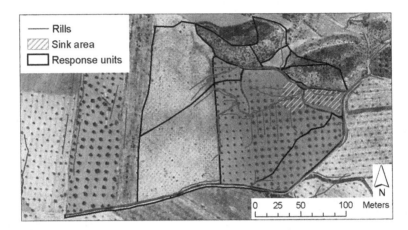

Fig. 4.2 Pathways of runoff and hydrological response units in Cárcavo (After Meerkerk et al. 2008)

spatial variability in soil moisture within the orchard. For the simulation years, there were 1–7 runoff events per year. Although runoff coefficients are low (1–5 %), runoff can still make an important difference in the spatial variability of soil moisture, tree growth and water availability for cover crops.

The evaporation loss from a 15-cm plough layer during autumn and winter, ranges from 100 to 155 mm, and varies little between the different soils. The evapotranspiration of a shallow rooting cover crop would use a similar amount of water. This means that there are opportunities for applying cover crops without a negative effect on the growth and yield of the tree crop. The amounts of runoff produced on the marl soils are relatively small.

This was studied in more detail for one almond field in Cárcavo, in order to determine whether the infiltration of runoff in depressions and thalwegs makes a significant difference in terms of water availability and vegetation growth (Fig. 4.2). The model results show that on an annual basis, the infiltration in the sink area within the field is 95–385 mm higher compared to the slope position (Table 4.1). An analysis of the tree trunk diameter of the trees in the sink area shows that the extra infiltration has a positive effect on tree vigour. The trunk basal area was 12 % larger at the age of 4 years ($p < 0.03$). This result shows that specific landscape positions may provide increased water availability and a higher potential for the growth of cover crops without competition with the tree crop in dry conditions.

4.2.2   Semi-Natural, Abandoned and Reforested Lands

Field monitoring of occurrence of runoff in different locations made it clear that vegetation has a significant effect on the pathways of water and sediment. The infiltration under vegetation patches is much higher and deeper, compared to bare

Table 4.1 Water balance of a 15 cm plough layer at different topographical positions. The reference refers to a position where run-on and run-off are equal (After Meerkerk et al. 2008)

	Year	Qon	I	Qoff	D	E	dS
Slope position:	Dry	–	166	3	0	166	0
	Average	–	285	2	39	245	1
	Wet	–	416	13	117	298	1
Sink position:	Dry	92	261	11	84	177	0
	Average	61	348	0	102	245	1
	Wet	372	801	189	476	324	1
Reference:	Dry	–	169	–	0	169	0
	Average	–	287	–	41	245	1
	Wet	–	429	–	130	299	0

Table 4.2 Reaction of runoff indicators after nine rainfall events

ID	Location	Vegetation cover (%)	Runoff occurrence
1	On 70 % south slope	50	4
2	At end of large gully	80	4.5
3	On 50 % south slope	15	6
4	At end of large gully	40	6.5
5	At end of small low gully in terrace wall	70	1.5
6	Before gully head	20	4.5
7	At end of large gully	30	7
8	At 45 % north slope	30	7
9	At 30 % north slope	90	3
10	Before gully head	55	3
11	Before gully head in terrace wall	40	6
12	At end of gully in the terrace wall	95	1

patches and also the temperature under vegetation is more regulated and the daily variation is less. The infiltration and sedimentation behind vegetation is also clear from the runoff indicators (Table 4.2). The occurrence of runoff for similar locations is more frequent when the vegetation cover is lower, e.g. runoff indicator 1 versus 3 or runoff indicator 2 versus 7. Threshold for runoff occurrence can be calculated by comparing the occurrence of runoff with the amount and intensity of rainfall.

Water repellency is associated with organic matter and vegetation. It was measured in the field using drop tests and time for penetration of water. Figure 4.3 shows the relationship between soil moisture, plant species and water repellency for *Rosmarinus* and *Anthyllis*. *Stipa* litter was found to be far less water repellent (not shown in figure). Figure 4.4 shows the relationship between accumulated organic matter and water repellency for the studied vegetation. For restoration purposes it should be taken into account that soils and accumulated organic matter with higher water repellency will increase the amount of local runoff, and may generate unwanted connectivity in the hydrological system. This may especially be a problem after prolonged dry periods with sudden high intensity rainfall.

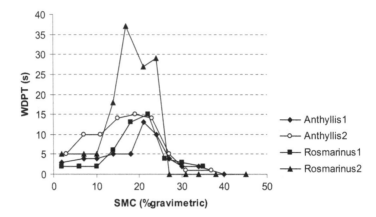

Fig. 4.3 Distribution of crust water repellency for two plant species in a wetting range. *SMC* soil moisture content, *WDPT* water drop penetration time. Each data point is the average WDPT of five droplets (Verheijen and Cammeraat 2007). Copyright © 2007 John Wiley & Sons, Ltd

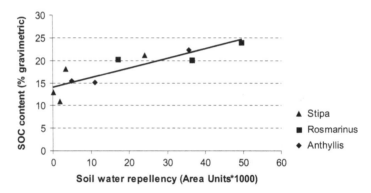

Fig. 4.4 The correlation of water repellency with SOC content of the mulch samples; r2 = 0.74 and p = 0.01. Insufficient mulch sample for SOC analysis was available for one Stipa, two Rosmarinus and two Anthyllis plants (Verheijen and Cammeraat 2007). Copyright © 2007 John Wiley & Sons, Ltd

The effect of vegetation on *soil crusts* was assessed by sampling thin sections of crusts on marls and calcretes with low organic carbon contents from fields that were abandoned and from semi-natural vegetation. The soil crusts were studied with respect to: (a) effects of material, (b) effects of geomorphological processes and (c) interaction with plants. The development of soil crusts was also studied in the laboratory under simulated rainfall.

The results from the thin section analysis showed that the crust properties are linked to their position in relation to plants and that these can explain hydrological behaviour. Figure 4.5 shows a clear contrast of soil surface properties up and down slope of a *Stipa tenacissima* tussock. The left photo shows a sedimentation crust with many vesicles and some organic matter accumulation, whereas the right

Fig. 4.5 Comparison between a sedimentation crust upslope a *Stipa tenacissima* tussock, and an erosion crust down slope a *Stipa* tenacissima tussock

photograph shows the badly developed soil surface with the marl substratum very close to the surface, as soil surface material has been washed away. This consequently has very important repercussions for runoff generation at the fine scale for areas such as described in Fig. 2.2 (Chap. 2). Figure 4.6 shows a multiple deposition crust under a fermentation layer on a reforestation terrace in Cretaceous marls. It also shows the change of runoff and sedimentation regime after the growth of the vegetation as the original crust prior to planting also is visible. This shows that overland flow frequently occurs, and that it is infiltrating for most of the events occurring. However, the terrace rim also showed serious signs of erosion from high magnitude events. Furthermore, crust evolution simulation revealed that sieving crusts, as present in the study area, can be created within 4 years. This was concluded from thin sections made from different stages of development after rainfall simulations in the laboratory and from sampled field sieving crusts that had developed after 40 years of abandonment in the same material.

4.3   Role of Plants in Reducing Concentrated Flow Erosion Rates

Within the RECONDES project detailed investigations were carried out into the effects of roots of Mediterranean plant species on soil resistance to concentrated flow erosion. Roots are the hidden half of plants that play a key role in rill-gully erosion and shallow mass movements (Vannoppen et al. 2015). Research methods

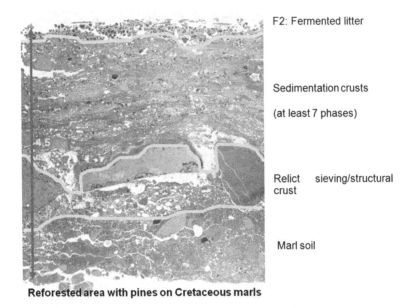

Fig. 4.6 Multiple sedimentation crusts on a reforestation terrace planted with *Pinus halepensis*

included laboratory flume experiments, fieldwork, and laboratory tensile and bend-ing strength measurements. Detailed methods and results have been published in a series of papers (see papers by De Baets et al.). They are summarised briefly here to illustrate the approaches.

4.3.1   Laboratory and Flume Experiments

The overall objective was to gain more insight into the influence of root architec-ture, soil and flow characteristics on the effects of plant roots in increasing the ero-sion resistance of topsoils during concentrated flow. More specific objectives were:

1. To assess the impact of root architecture (tap root systems vs. fine-branched root systems) on the erosion-reducing potential of roots during concentrated flow,
2. To assess the impact of soil texture (sandy loam vs. silt loam) on the potential of roots to increase the resistance of the topsoil against concentrated flow erosion,
3. To assess the impact of soil moisture content (wet vs. dry topsoil samples) on the erosion-reducing potential of roots during concentrated flow,
4. To assess the impact of flow shear stress on the erosion-reducing potential of roots during concentrated flow.
5. To determine the effect of the spatial organisation of plants on the erosion-reducing potential of roots during concentrated flow.

Six field plots (length=1.19 m, width=0.97 m, and soil depth=0.15 m for each) were established at Leuven University Campus on a sandy loam soil (8 % clay/< 2 μm, 36 % silt/50–2 μm and 56 % sand/50 μm–2 mm in year 1). In year 2 six plots were established at the same place on a silt loam soil (9 % clay/< 2 μm, 70 % silt/2–50 μm and 21 % sand/50 μm–2 mm). Treatments were: (1) bare; (2) grass (simulating fine-branched roots, low and high density drilling), and; (3) carrots (simulating taproots, low and high density drilling). For the assessment of the effect of spatial organisation of plants, treatments were: (1) bare; (2) grass randomly sown, and; (3) grass sown in rows with an inter-row distance of 5 cm. Differing variables between the samples tested are type of species, flow shear stress level, soil type and soil moisture conditions.

Laboratory experiments simulating concentrated flow were conducted with a flume similar to the one used by Poesen et al. (1999) (Fig. 4.7; length=2 m, width=0.098 m). The slope of the flume surface could be varied and clear tap water flow could be simulated at a known constant discharge. The erosion parameters calculated included relative soil detachment rate (*RSD*). *RSD* was calculated as the ratio between absolute soil detachment rate (*ASD*) for the root-permeated soil samples and the *ASD* for the bare topsoil samples, tested at the same time. *ASD* rate for each sample was calculated using the following equation:

$$ASD = (SC * Q) / A \qquad (4.2)$$

where *SC* is sediment concentration (kg l^{-1}), *Q* is flow discharge (l s^{-1}) and *A* is area of soil sample surface (m^2).

Fig. 4.7 Hydraulic flume at KU Leuven used to measure detachment rates from root permeated topsoil samples. A is the test section (length=38.8 cm, depth=8.8 cm, width=9 cm). *Arrows* indicate concentrated flow direction

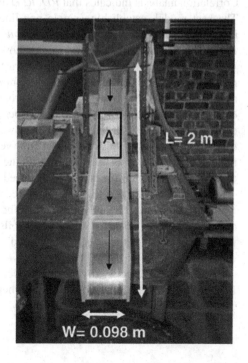

Assessed root parameters were root density (*RD*), root length density (*RLD*) and root diameter (*D*) because of their relevance for evaluating the effects of roots on erosion processes (Gyssels and Poesen 2003; Li et al. 1991; Mamo and Bubenzer 2001a, 2001b). *RD* (kg m^{-3}) in this study is expressed by the dry mass of the living roots divided by the volume of the root-permeated soil sample:

$$RD = \frac{M_D}{V} \qquad (4.3)$$

where M_D is dry living root mass (kg) and *V* is volume of the sample box (m^3).

RLD is the total length of the roots divided by the volume of the root-permeated soil sample (Smit et al. 2000):

$$RLD = \frac{L_R}{V} \qquad (4.4)$$

where L_R is length of the living roots (km).

In this study mean root diameter (*D*, mm) for each sample was also assessed to investigate the suitability of this root parameter for improving the prediction of the erosion-reducing effect of roots (De Baets et al. 2006, 2007a).

4.3.1.1   Results

Correlation analysis indicates that *RD*, *RLD* and *D* are well correlated with ln*RSD*. The best model, establishing single non-linear regression equations between all influencing variables and *RSD* *(relative soil detachment rate)*, was obtained with *RD* ($R^2 = 0.59$). Our experimental results show that roots can reduce erosion rates to very low values (for very low *RD*), in contrast to what was previously thought. The results of the laboratory experiments do confirm that the erosion-reducing effect of roots is dependent on root architectural properties. Root diameter (*D*) does influence the erosion-reducing potential of roots during concentrated flow. *RD* and *D* explain the observed variance in *RSD* very well. The erosion-reducing effect of *RD* decreases with increasing *D*. Carrots with very fine roots can reduce soil detachment in a similar way as grass roots. With increasing *D*, the erosion-reducing effect of carrot type roots becomes less pronounced compared to fine roots.

The results indicate a higher erosion-reducing potential of grass roots for silt loam soils compared to sandy loam soils. The erosion-reducing effect of both grass and carrot roots is larger for initially wet soils compared to dry soils, because slaking strongly reduces the effect of increased cohesion provided by roots (De Baets et al. 2007a).

Grasses are as effective in reducing concentrated flow erosion for low as for high flow shear stresses. Carrot roots on the other hand, are less effective in reducing erosion rates when high flow shear stresses are applied. For carrots *RSD* values higher than 1 can be observed, even when tested at a low flow shear stress. When tested at medium and high shear stresses, *RSD* values are often higher for samples with carrot roots compared to grass roots. For carrots growing in a sandy loam soil, significant differences (at the 5 % level) can be observed between shear stress levels, when the root effect is predicted with *RD*. This can be explained by the occurrence of local turbulence and vortex erosion scars around individual carrot roots, which form an obstacle to the flow and increase the detachment rate (De Baets et al. 2007a).

A metadata analysis showed that for fibrous root systems, models using root density and soil moisture information are capable to explain 79 % of the variation in relative erosion rates, whereas relative erosion rates for tap-root permeated topsoils remain difficult to predict with root architectural information only (De Baets and Poesen 2010).

Experimental results also showed that both a random orientation and rows perpendicular to the dominant flow direction are effective in reducing concentrated flow erosion rates by roots. Planting or sowing species in rows parallel to the flow direction does not offer a good protection to erosion by concentrated flow (De Baets et al. 2014).

4.3.2   Field Measurements

Root characteristics of typical Mediterranean plants were measured in the field and an empirical model, linking root characteristics (root density and root diameter) to relative soil detachment rate, was then used to predict the root effect on the resistance to concentrated flow erosion. Plants were sampled in three different habitats, i.e. the ephemeral river channels (Group 1), on abandoned croplands formed in Quaternary loams (Group 2) and in badland areas (semi-natural land consisting of steep incised slopes) formed in marls (Group 3). The root characteristics measured are root:shoot ratio, root density, root diameter and root length density.

The plants are carefully excavated in the field then, after the excavation, digital photos of the root systems are made to describe the root system. Height and diameter of the orthogonal projection of the above ground biomass are measured with a ruler. Then the root system is put on a horizontal plastic sheet (Fig. 4.8) and exposed in the same position as it is growing in the soil and roots are cut into soil depth classes of 10 cm starting for the bottom up to the top of the root system. Roots are collected in a plastic bag per soil depth class and per species. In the laboratory roots from each bag were selected in four diameter classes: <2 mm, 2–5 mm, 5–10 mm, >10 mm. For each diameter class and each depth class root dry mass was assessed. Therefore the roots were put into the oven for 24 h at 60–65 °C (Smit et al. 2000).

Fig. 4.8 Five excavated *Anthyllis* root systems, Cárcavo, Spain (Photo: S. De Baets, January 2005)

Root dry weight per individual and per depth class is then obtained by calculating an average value. Roots from each depth class are also exposed to take digital photographs from which total length was assessed using a digitisation software program (Mapinfo Professional 6.0). For an individual plant species root length per depth class is the average value of the root length of all the plants measured. The total wet above-ground biomass was weighed in total in the field with a field balance. A sample of this biomass was taken to the lab for calculation of moisture content and mean total dry mass of the shoots per individual plant.

To assess the resistance of topsoils reinforced with Mediterranean plant roots to erosion by concentrated flow, the empirical relationships, established for carrots (*Nandor F1 hybride* seeds), weeds (e.g. *Trifolium Repens* L.) and grass (27 % of *Lolium perenne*, 21 % of *Lolium perenne*, 12 % of *Festuca rubra* and 40 % of *Festuca arundinacea*) roots during simulated flow experiments in the laboratory on a saturated silt loam soil, were used. These experiments resulted in a negative exponential relationship between relative soil detachment rate (RSD) and RD as a function of root diameter. Different equations could be established to describe the erosion-reducing effect for plant roots of different root diameter classes, whereby the erosion-reducing effect of plant roots decreases with increasing root diameter.

The values of the RSD for the different species within the first 10 cm of topsoil are listed in Table 4.3. This list has been ranked according to the predicted RSD values for the topsoil (0–0.10 m) in order to select species according to their erosion reduction potential. It can be observed that the roots from grasses (family of the *Poaceae*) such as *Helictotrichon filifolium, Piptatherum miliaceum, Juncus acutus, Avenula bromoides, Lygeum spartum* and *Brachypodium retusum* have a very high

Table 4.3 List of species and the potential of their root system for increasing the erosion resistance of topsoils to concentrated flow erosion (De Baets et al. 2007b)

Name of the species	RSD (0–10 cm topsoil)	Erosion reducing Potential
Avenula bromoides	$0.3.10^{-12}$	Very high
Juncus acutus	$2.72.10^{-8}$	Very high
Lygeum spartum	$2.41.10^{-7}$	Very high
Helictotrichon filifolium	$1.61.10^{-6}$	Very high
Plantago albicans	1.10^{-5}	Very high
Brachypodium retusum	8.10^{-4}	Very high
Anthyllis cytisoides	$2.29.10^{-3}$	Very high
Piptatherum miliaceum	0.01	Very high
Tamarix canariensis	0.01	Very high
Stipa tenacissima	0.03	High
Retama sphaerocarpa	0.03	High
Salsola genistoides	0.03	High
Artemisia barrelieri	0.07	High
Dorycnium pentaphyllum	0.11	Medium
Rosmarinus officinalis	0.15	Medium
Atriplex halimus	0.18	Medium
Nerium oleander	0.19	Medium
Dittrichia viscosa	0.19	Medium
Fumana thymifolia	0.25	Low
Thymus zygis	0.32	Low
Teucrium capitatum	0.32	Low
Limonium supinum	0.37	Low
Ononis tridentata	0.45	Low
Thymelaea hirsuta	0.5	Very low
Phragmites australis	0.6	Very low
Bromus rubens	0.71	Very low

RSD relative soil detachment rate for the first 10 cm of the topsoil ($0 =$ very high erosion resistance, $1 =$ very low erosion resistance), $0 < RSD < 0.01 =$ very high erosion-reducing potential, $0.01 < RSD < 0.10 =$ high erosion-reducing potential, $0.10 < RSD < 0.25 =$ medium erosion-reducing potential, $0.25 < RSD < 0.50 =$ low erosion-reducing potential, $RSD > 0.50 =$ very low erosion-reducing potential

concentrated flow erosion-reducing potential ($0 < RSD < 0.01$) for the 0–0.10 m thick topsoil. This can be attributed to the high density of fine roots in the topsoil for these species. These grasses only protect the 0–0.20 m thick topsoil. Moreover, comparison of the slope values of the relationship between RSD and soil depth (for the top 0–0.20 m) indicated that the erosion-reducing effect of grasses diminished very rapidly with increasing soil depth compared to the other studied plant species.

4.3.3   Assessment of Stem Density and Trapping Effectiveness

In this study, stem density (SD) was measured for each individual plant species ($n=5$ plants per species). The unit area equals the soil surface occupied by the vertical projection of the above-ground biomass and is thus species dependent. The stems were assumed to have a circular cross-section. Stem density for shrubs (SD, m^2 m^{-2}) is calculated as:

$$\text{For Shrubs:} \quad SD = \frac{\sum \pi (d_i/2)^2}{\pi (D_P/2)^2} \tag{4.5}$$

where d_i is the diameter of each stem (m) and D_P is the mean diameter of the vertical orthogonal projection of the above-ground biomass (m).

For grasses not all stem diameters were measured separately. Only a representative horizontal area (ca. 5 cm^2) of a section of grasses (S_s, m^2), the corresponding number of stems in this section (n_s) and their mean diameter were measured (d). Additionally the fraction of this area relative to the total surface occupied by vertical projection of the above-ground biomass was assessed to calculate total stem density for grasses (Eq. 4.6).

$$\text{For grasses:} \quad SD = \frac{S_{tot}}{S_s} * \frac{n_s \cdot \pi \left(\frac{d}{2}\right)^2}{S_s} \tag{4.6}$$

where n_s is the number of stems in a horizontal section with area S_s, d is mean stem diameter, S_{tot} (m^2) is the total surface occupied by the vertical projection of the above-ground biomass.

A measure of the sediment and organic debris trapping effectiveness (TE, m m^{-1}) of plants is the ratio of the diameter (d_i) of the horizontally projected stems on a line perpendicular to the dominant flow direction and the maximum length of this perpendicular line (L_{tot}) defined by the vertical projection of the above-ground biomass, i.e.

$$\text{For shrubs:} \quad TE = \frac{\sum d_i}{L_{tot}} \tag{4.7}$$

where TE is trapping effectiveness, d_i is stem diameter and L_{tot} is the length determined by the projection of the above-ground biomass, in a direction perpendicular to the dominant flow direction (assessed topographically). Bold lines indicate the vertical projection of maximum extent of above-ground biomass. d_i values represent the width of the horizontal projection of all plant stems on L_{tot}.

Stem density (SD) was calculated according to Eqs. 4.5 and 4.6 as a measure for the capacity of plant stems to trap detached sediment and organic residues arriving from different directions. SD ranges from 0.04 to 12.61 %. The two herbs *Plantago albicans* and *Limonium supinum* show the highest stem density measurements. These are small plant species with a good ground cover. Some grasses like *Stipa tenacissima* and *Helictotrichon filifolium* and the studied reeds *Juncus acutus* and *Phragmites australis* also show high stem densities. Other grasses like *Avenula bromoides* and *Lygeum spartum* do not have such high stem densities as expected. This can be explained by the angle of the stems, whereby the total surface under the crown is higher as compared to other grasses.

Trapping effectiveness (TE) of shrubs was measured according to Eq. 4.7 and similarly for herbs and grasses along a line orthogonal to the dominant flow direction, as a measure for the sediment trapping effectiveness assuming that sediment originates from one direction. TE ranges between 3.3 and 35.4 %. The two herbs *Plantago albicans* and *Limonium supinum*, as well as most of the grasses (except *Lygeum spartum*) and the reeds *Juncus acutus* and *Phragmites australis* are shown to be very effective for trapping sediment and organic residues. Also some shrubs like *Rosmarinus officinalis* and *Nerium oleander* have a high obstruction capacity.

More information on stem density and trapping effectiveness values and assessment methods can be found in De Baets et al. (2009).

4.3.4   Laboratory Root Tensile Strength and Stem Bending Tests

Using the same apparatus as used for the tensile strength tests, stem bending tests were also performed to assess modulus of elasticity for individual stems of 25 typical Mediterranean plant species in the laboratory. Root tensile strength (T_r, MPa) tests were conducted in the laboratory with a UTS testing apparatus (Fig. 4.9). The following formula was used to calculate T_r (Bischetti et al. 2005b):

$$T_r = \frac{F_{max}}{\pi \left(\dfrac{D^2}{4} \right)} \tag{4.8}$$

where F_{max} is the maximum force (N) needed to break the root and D is the mean root diameter (mm) before stretching.

Modulus of elasticity (E_{mod}) is a material characteristic and can be calculated from the rigidity divided by the second moment of inertia ($I = \pi r^4/4$ for circular stems). For each species, ten samples of 15 cm long stems were tested. Different stem diameters were tested. As reported in many studies (e.g. Bischetti et al. 2005b; Mattia et al. 2005; Norris 2005; Operstein and Frydman 2000; Tosi 2007) root tensile strength (Tr) decreases with increasing root diameter (D), following a power

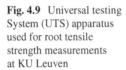 **Fig. 4.9** Universal testing System (UTS) apparatus used for root tensile strength measurements at KU Leuven

law equation of the form $Y = a \, X^b$. The maximum recorded root tensile strength values amount to 303 mPa for *Nerium oleander* (D = 0.09 mm) and 267.5 mPa for *Tamarix canariensis* (D = 0.1 mm) (De Baets et al. 2008). Root tensile strength values are in the same range of root tensile strength values reported by Bischetti et al. (2005a) for forest species in Northern Italy. Mean root tensile strength does not differ significantly (p = 0.12) according to vegetation type (De Baets et al. 2008).

The results from a bending test on the stems show that grasses and fragile shrubs or herbs like *Fumana* and *Limonium* have a high rigidity per unit stem cross-sectional area. The product MEI of stem density (M), modulus of elasticity (E) and moment of inertia (I) combines bending properties with stem density and stem morphology. The results in Table 4.4 show that the shrubs and trees are the most resistant and the grasses and some small shrubs the least resistant to bending simulated under flow shear forces. The reeds show intermediate values.

4.3.5   Synthesis

Based on the various properties measured of the different plant species, a cluster analysis was performed. This analysis resulted in eight clusters, whereby all plants belonging to a cluster show similar performance on the criteria used to select the suitability of species for erosion control (Table 4.5). For more detailed information on the criteria and methods used to select suitable species for rill and gully erosion control reference is made to De Baets et al. (2009).

Table 4.4 MEI values for Mediterranean plant species

Plant name	Vegetation type	MEI (N)
Avenula bromoides	Grass	0.45
Helictotrichon filifolium	Grass	3.62
Fumana thymifolia	Shrub	0.64
Limonium supinum	Herb	0.25
Teucrium capitatum	Shrub	0.62
Lygeum spartum	Grass	0.41
Stipa tenacissima	Grass	0.32
Brachypodium retusum	Grass	0.05
Artemisia barrelieri	Shrub	2.70
Piptatherum miliaceum	Grass	2.09
Juncus acutus	Reed	1.77
Dittrichia viscosa	Shrub	13.58
Dorycnium pentaphyllum	Shrub	22.27
Phragmites australis	Reed	5.54
Anthyllis cytisoides	Shrub	8.56
Thymelaea hirsuta	Shrub	17.04
Atriplex halimus	Shrub	54.16
Thymus zygis	Shrub	11.96
Tamarix canariensis	Tree	48.48
Nerium oleander	Shrub	25.17
Rosmarinus officinalis	Shrub	159.80
Salsola genistoides	Shrub	42.17
Retama sphaerocarpa	Shrub	27.51
Ononis tridentata	Shrub	97.54

4.4 Effects of Vegetation in Channels

4.4.1 Roughness and Hydraulics

Determining the effect that different plant assemblages have on flow hydraulics is key to evaluating the potential for vegetation to withstand erosive forces of floods, increase the overall resistance of the channel to erosion and reduce the connectivity of sediments along river channels. In order to calculate hydraulics surveyed cross-sections and estimated Manning's n values for specified flow stages are entered as input into software package WinXSPRO (Hardy et al. 2005) and the velocity and discharge is computed. The output generated by WinXSPRO can then be used to calculate other hydraulic values such as unit power, shear stress and shear velocity. The advantage of WinXSPRO is that the cross-sections can be divided into subsections and different Manning's n values applied (Sandercock and Hooke 2010). Thus differentiation due to different zones of vegetation can be incorporated. The values of n were assigned according to guidelines in Arcement and Schneider (1989).

Table 4.5 Plant species grouped in eight clusters according to their scoring for the four main requirements, i.e. (1) the potential to prevent incision by concentrated flow erosion, (2) the potential to improve slope stability, (3) the potential to resist bending by water flow and (4) the ability to trap sediments and organic debris (De Baets et al. 2009 Earth Surface Processes and Landforms 34: 1374–1392)

Cluster	Plant species name	Cluster description
1	*Fumana thymifolia* *Teucrium capitatum*	low resistance to erosion, low trapping effectiveness, not resistant to removal, no potential to improve slope stability
2	*Nerium oleander* *Rosmarinus officinalis*	low potential for slope stabilisation medium potential to prevent erosion by concentrated runoff high resistance to bending, medium trapping effectiveness
3	*Anthyllis cytisoides* *Retama sphaerocarpa* *Salsola genistoides* *Tamarix canariensis* *Atriplex halimus*	high potential for slope stabilization, very resistant to removal, low trapping effectiveness and medium to high potential to prevent concentrated flow erosion
4	*Thymus zygis* *Artemisia barrelieri* *Lygeum spartum* *Avenula bromoides* *Piptatherum miliaceum*	medium to high potential to prevent erosion, low potential for slope stabilisation medium trapping effectiveness and low resistance to removal
5	*Stipa tenacissima* *Thymelaea hirsuta* *Dittrichia viscosa* *Ononis tridentata* *Dorycnum pentaphyllum*	medium potential for slope stabilisation, low potential to prevent incision medium to high resistance to removal, low trapping effectiveness
6	*Plantago albicans* *Limonium supinum* *Helictotrichon filifolium* *Brachypodium retusum*	medium to high potential to prevent concentrated flow erosion, easy to remove low potential for slope stabilization and high trapping effectiveness
7	*Juncus acutus*	high potential to prevent erosion, high trapping effectiveness and medium resistant to removal
8	*Phragmites australis*	medium resistant to removal, high trapping effectiveness, medium to low potential to prevent erosion

Tables of component Manning's n values have been compiled based on extensive work on channels in Arizona where roughness values have been measured (Aldridge and Garrett 1973; Arcement and Schneider 1989) and verified (see O'Day and Phillips 2000; Phillips and Hjalmarson 1994; Phillips and Ingersoll 1998; Phillips et al. 1998), and which have similarities to those which are being studied in Southeast Spain. The hydraulics of flows have been calculated at selected sites for minor, moderate and high discharge events along two channels in Southeast Spain, Cárcavo (2, 10 and 40 m^3 s^{-1}) and Torrealvilla (10, 100 and 200 m^3 s^{-1}). These discharges were selected as having ecological and geomorphological significance. The low flow fills the thalweg/inner channel, the medium flood most closely resembles that

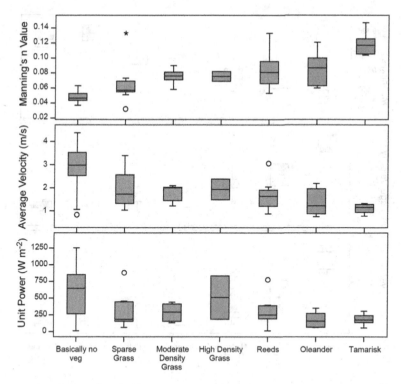

Fig. 4.10 Output from hydraulic computations for high flow (40 m³ s⁻¹) along Cárcavo showing the influence that Manning's n value for different vegetation types has on average velocity and unit power (After Sandercock and Hooke 2010)

of a bankfull and the high flow is at least twice the magnitude of the medium flow. Very sparse information is published or available on the frequency and magnitude of flows but monitoring of flows within the Guadalentín Basin over the previous 10 years also informed selection of these flows (Hooke 2007a).

As is to be expected, as Manning's n values increase there is a reduction in the velocity and unit power of flows, this reflecting in part the effect that vegetation has in increasing the roughness of the channel. This is demonstrated most clearly in Figs. 4.10 and 4.11 which show the output of the hydraulic computations for the high flood at Cárcavo and Torrealvilla sites respectively. Highest velocities are associated with grasses which have low Manning's n values (they are also positioned low in the channel). Comparably lower velocities are also associated with Oleander and Tamarisk, this attributed to the high Manning' n values but also their tendency to establish on more elevated surfaces. Reeds are also associated with lower flow velocities, this in part is a function of the very low gradients where these plants establish.

To assess the resistance of different plants to erosion, these models need to be informed by field surveys, where flood effects on vegetation have been documented.

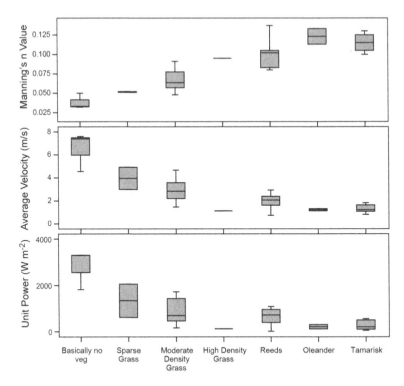

Fig. 4.11 Output of hydraulic computations for high flow (200 m³ s⁻¹) along Torrealvilla showing influence that Manning's n value has on average velocity and unit power (After Sandercock and Hooke 2010)

For this, we drew upon the impact that floods in September 1997 (Hooke and Mant 2000) and October 2003 within the Guadalentín Basin (Salada) and November 2006 in Cárcavo have had on vegetation and channel morphology. The hydraulics of floods and forces acting on vegetation were calculated using detailed cross-sections and flood debris lines. The effects of these floods on vegetation provide the basis for assessing the thresholds of forces for plant damage and destruction. This information can then be fed back into the models, to predict what would be the likely response of the vegetation and channel morphology to a particular flood. The results in relation to a range of species found in the channels and on floodplains are presented in Fig. 4.12, documenting cases where vegetation suffered no change, battered, swept over, flattened, mortality or removed in relation to calculated shear stress and velocity values. Instances of mortality or removal of vegetation are rare, though some data do exist.

Of the two dominant species of grasses that are found in these channels, *Piptatherum miliaceum* and *Lygeum spartum*, the latter demonstrates greater resistance to flows. *Piptatherum miliaceum* are swept over and flattened by flows with shear stresses of <200 N m⁻² and velocities of 1.7 m s⁻¹. Some cases of removal

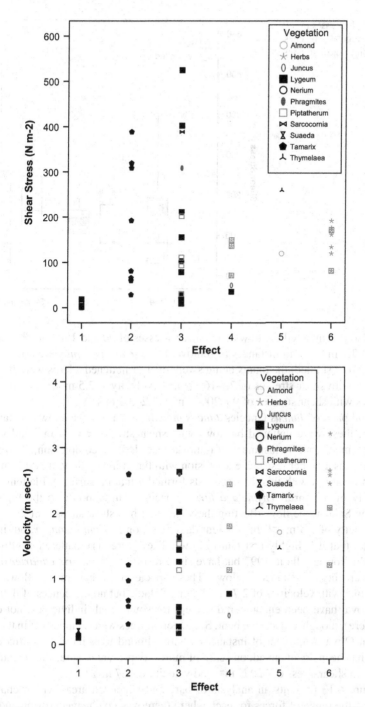

Fig. 4.12 Summary of effects on vegetation for flows in Torrealvilla (September 1997), Salada (November 2004) and Cárcavo (November 2006) in relation to calculated shear stresses and velocity. 1 = no change, 2 = battered, 3 = swept over, 4 = flattened, 5 = mortality, 6 = removed

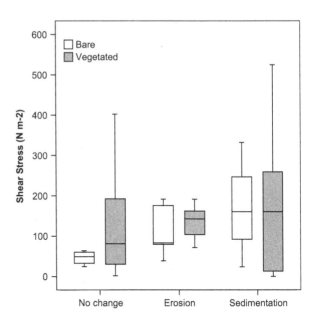

Fig. 4.13 Processes in bare and vegetated patches in relation to shear stresses of flow

were also documented for flows with shear stresses of 90 and 180 N m⁻² (Velocity 1.2 and 2.1 m s⁻¹). No instances of removal of *Lygeum spartum* were documented for the flows studied, the plants being swept over or flattened. Herbs were flattened by flows with shear stresses of 70–160 N m⁻² (velocity 1–2.5 m s⁻¹) and removed by flows with shear stresses of 90–200 N m⁻² (1.2–3.2 m s⁻¹).

Aerial parts of *Juncus* (species *Juncus maritimus*), also have a low resistance to erosion: these were flattened by flows with shear stresses < 50 N m⁻² and velocities < 0.5 m s⁻¹ (but no instances of removal recorded). In comparison, *Phragmites australis* has a higher resistance to erosion with flattening by flow of 300 N m⁻² (no erosion occurred with this flow as reeds formed a mat on surface). Flattening and mortality of the shrub *Thymelaea hirsuta* occurred in response to the high flow event on Salada; the flow affecting the vegetation had shear stresses of 250 N m⁻² and velocity of 1.5 m s⁻¹. Insufficient data exists on *Nerium oleander*, but indications are that it is highly resistant to flows. They were very battered at Oliva site along the Torrealvilla in 1997 but have since resprouted. *Tamarix canariensis* also has a very high resistance to flows. These species were battered by flows up to 400 N m⁻² with velocities of 2.7 m s⁻¹ along Salada, but no incidences of flattening or removal have been encountered (except for some small individuals, not calculated here), though they can be bent. Some poplar trees were snapped off in the 1997 flood at Oliva. A number of instances where Almond trees have been affected by floods has been documented, one case of mortality presented here, was a result of a flow with shear stress of 120 N m⁻² and velocity of 1.7 m s⁻¹.

Figure 4.13 presents an analysis of bare and vegetated areas within channels, showing the range of forces for each where there was no change, erosion and sedimentation. The effect of vegetation in increasing the resistance of the channel to erosion is clear when comparing the distribution of forces for vegetated patches against bare patches. Within vegetated patches, flows up to 400 N m^{-2} with velocities of 3.2 m s^{-1} were not capable of causing erosion, whilst bare areas had a much lower upper limit of flows which caused erosion (shear stresses < 70 N m^{-2} and velocities < 1.7 m s^{-1}).

Minor floods have relatively little impact on vegetation and channel form, patches of reeds may be swept over, with some minor incision being restricted to the thalweg (such as the November 2006 event). The overall impact of this flood is positive, contributing water to growth of plants and refilling local aquifers. A moderate flood (such as the September 1997 event) results in greater incision and widening of the channel and exceeds the threshold for the removal of herbs, and battering of vegetation. These events have a negative impact on vegetation, however, recovery can be fairly rapid. A major flood (such as the October 2003 event) is likely to cause significant morphological changes, removal of herbs and flattening of vegetation. An extreme flood, such as that which occurred in the region in October 1973 and in September 2012 with flow depths up to 5 m is likely to cause major channel adjustments, with erosion and widening of the channel, massive scour and headcut formation and extension. All vegetation may be removed, although *T. canariensis* may withstand such an event. Subsequent events may have greater geomorphological effect due to the degraded condition of the channel.

4.5 Summary

Studying the effect plants have on processes has been a major component of the project and these have been reported on in this chapter.

A climatic threshold for cover crops was identified, based on the relation between tree density, projected canopy cover and climate. As a climatic indicator the Humidity Index (*HI*) which is the ratio between annual precipitation and evapotranspiration was used. For locations where *HI* > 0.6, the growth of olive trees is no longer limited by water availability and there is a surplus of water that could be used to grow cover crops. Below this threshold, the use of cover crops must be restricted in space and time to avoid competition for water. Field surveys of roots showed that deep rooted almond crops and shallow rooted cover crops derive their water from different parts of the soil. The results of further modelling of water balance in the plough layer showed that more water is available for cover crops in orchards on marly soils than in orchards on stony soils, the difference being related to the deeper infiltration of rain on the stony soils. The results of this work indicate that there are opportunities for applying cover crops without a negative effect on the growth and yield of the tree crop. In addition, results from water budget modelling for one almond field in Cárcavo indicated that there was greater water availability for cover crops in depressions.

Vegetation has a significant influence on hydrological pathways as shown in the work on semi-natural and abandoned lands. As vegetation develops it modifies pathways of water and sediments, with higher infiltration under vegetation patches and sedimentation behind vegetation. Water repellency also plays an important role in the local redistribution of water over soils: accumulated organic matter and water repellency may increase runoff and generate unwanted connectivity. The formation of soil crusts in bare areas after land abandonment also reduces the infiltration capacity of the soil resulting in increases in runoff and erosion.

A large amount of data has been collected on the above and below ground attributes of plants that are important in reducing erosion, From the various experiments and field measurements, both root and shoot properties have then been combined to assess the suitability of plants for erosional control in three different areas prone to water erosion: channels, abandoned croplands and steep badland slopes. Grasses like *Helictotrichon filifolium, Piptatherum miliaceum, Avenula bromoides, Lygeum spartum* and *Brachypodium retusum* have the highest potential to reduce soil erosion rates, but shrubs such as *Anthyllis cytisoides* and the tree *Tamarix canariensis* are also effective in reducing erosion rates. Within channels *Juncus acutus* and *Tamarix canariensis* have the highest potential to reduce erosion. *Salsola genistoides* is the most suitable species to control gully erosion on steep slopes, but also the grasses *Helictotrichon filifolium* and *Stipa tenacissima* can be planted to increase the resistance to erosion. On Abandoned fields, *Rosmarinus* and *Plantago albicans* are the most effective species in reducing concentrated flow erosion rates.

From measurements of plant properties and of effects of flows in channels, effectiveness of plants to reduce erosion and increase sedimentation in channels has been assessed. Grasses such as *Lygeum spartum*, the shrub species *Nerium oleander* and Tree species *Tamarix canariensis* are found to be highly resistant to removal. Of particular interest is *Lygeum spartum*, having an affinity for fine sediments, and its ability to trap further fines and form vegetated mounds. Spatial comparisons show that the effect vegetation has in trapping sediments is dependent on its location in the channel network.

Chapter 5
Synthesis and Application of Spatial Strategies for Use of Vegetation to Minimise Connectivity

Janet Hooke, Peter Sandercock, Gonzalo Barberá, Victor Castillo,
L.H. Cammeraat, Sarah De Baets, Jean Poesen, Dino Torri,
and Bas van Wesemael

Abstract The knowledge that has been acquired in the project RECONDES on critical conditions necessary for plants and on the occurrence of such conditions in the landscape is used in combination with the analysis of processes to develop strategies that could be applied at critical points and locations, identified by the connectivity mapping, to produce greatest effectiveness of the vegetation measures. This is achieved at the plot and land unit scale based on measurements of plant conditions, and at the catchment scale with the analysis of vegetation cover and conditions, both scales involving identification of erosion hotspots from connectivity mapping and modelling. These results have provided the framework for recommendations on spatial strategies and targeting of revegetation and restoration. The analysis of the effectiveness of different types of plants and species is used to select appropriate plants for different locations in the landscape. This has informed the development of practical guidelines produced for use by land managers and advisors. The research was developed in a Mediterranean environment but has wider applicability to drylands prone to erosion by water.

Electronic supplementary material The online version of this chapter (doi: 10.1007/978-3-319-44451-2_5) contains supplementary material, which is available to authorized users.

J. Hooke (✉)
Department of Geography and Planning, School of Environmental Sciences,
University of Liverpool, Roxby Building, L69 7ZT Liverpool, UK
e-mail: janet.hooke@liverpool.ac.uk

P. Sandercock
Jacobs, 80A Mitchell St, PO Box 952, Bendigo, VIC, Australia
e-mail: Peter.Sandercock@jacobs.com

G. Barberá • V. Castillo
Centro de Edafologia y Biologia Aplicada del Segura (CEBAS),
Department of Soil and Water Conservation, Campus Universitario
de Espinardo, PO Box 164, 30100 Murcia, Spain

L.H. Cammeraat
Instituut voor Biodiversiteit en Ecosysteem Dynamica (IBED)
Earth Surface Science, Universiteit van Amsterdam, Science Park 904,
1098 XH Amsterdam, The Netherlands

Keywords Erosion control • Restoration strategies • Plant species selection •
Connectivity minimisation • Land restoration guidelines • Desertification mitigation

5.1 Introduction

The knowledge that has been acquired on critical conditions necessary for plants
and on the occurrence of such conditions in the landscape is used in combination
with the analysis of processes to develop strategies that could be applied at critical
points and locations, identified by the connectivity mapping, to produce greatest
effectiveness of the vegetation measures. This is achieved at a number of scales: (1)
at the plot and land unit scale based on measurements of plant conditions and identification of hotspots from detailed connectivity maps; and (2) at the catchment
scale with the analysis of vegetation cover and conditions and identification of
hotspots from the connectivity modelling. The outcomes of the work undertaken at
each of these scales have been incorporated into the model EUROSEM, which has
also been used to investigate the effect different measures have on connectivity. The
analysis of the effectiveness of different types of plants and species is used to select
appropriate plants for different locations in the landscape. This has also informed
the development of practical guidelines (http://www.port.ac.uk/research/recondes/
practicalguidelines/). *A copy of these guidelines can be found in the Extra Material
associated with this volume.*

The following sections detail the outcomes of the analyses and the synthesis of
the vegetation and connectivity components of the research to produce the spatial
strategies for each type of land unit and at each scale. In each section we consider
the criteria for selection of species, the conditions and constraints on implementation in that type of location, and the optimisation of location of the vegetation. The
desirable characteristics of plants for greatest effectiveness in reducing erosion and
sediment flux or encouraging infiltration of water were listed in Table 1.2 (Chap. 1)
but are presented here again for reference (Table 5.1).

S. De Baets
College of Life and Environmental Sciences, Department of Geography,
University of Exeter, Rennes Drive, EX4 4RJ Exeter, UK

J. Poesen
Division of Geography and Tourism, Department of Earth and Environmental Sciences,
KU Leuven, Celestijnenlaan 200E, 3001 Heverlee, Belgium

D. Torri
Consiglio Nazionale Della Ricerche – Istituto di Ricerca per la Protezione Idrogeologica
(CNR-IRPI), Via Madonna Alta 126, Perugia, Italy

B. van Wesemael
Earth and Life Institute, Université catholique de Louvain (UCL),
Place Louis Pasteur 3, 1348 Louvain-la-Neuve, Belgium

Table 5.1 Desirable plant characteristics for erosion control	Properties
	Native species
	Germinates and/or propagates easily
	Rapid growth rate
	Perennial or persistent
	Ability to grow in a range of substrates
	Drought tolerant
	Produces a dense root system
	Has a high threshold to withstand forces of water flow
	Ability to trap water and sediments

5.2 Application at Hierarchical Scales

5.2.1 Land Unit Scale

5.2.1.1 Reforested Lands

Extensive reforestation works using terracing to reduce runoff and improve soil water availability for plants have been carried out in Spain and elsewhere for more than three decades as the main method to control soil erosion in degraded areas. As a consequence, large areas of the Mediterranean landscape have been topographically altered and covered by a homogeneous, even-aged, and single-species planted forest, primarily *Pinus halepensis*. Landscape remodelling was undertaken using heavily mechanized techniques, such as bulldozer terracing and subsoiling, but the altered topography and vegetation cover have given rise to degradation processes which it is necessary to act upon. Terracing involves, to varying degrees, a modification of topography that can affect landscape connectivity. Defective terraces, which are not perpendicular to slope, act as fast runoff pathways. Local default in terraces seems to increase the hydrological connectivity. Finally, higher connectivity is favoring the collapse of old reforestation terraces and the migration upstream of the drainage network (Chap. 2).

Most reforestation works were carried out without considering the inherent spatial heterogeneity of conditions within Mediterranean landscapes. Thus, success of reforestations has been very varied depending on mesoscale environmental characteristics (climatic conditions, lithology, soil type and local water balance) and soil preparation techniques. Dense stands of planted trees exist alongside almost bare patches where only a few stunted plants have survived. A medium term survival rate of 50 % survival rate has been reported in studies on effectiveness of *P. halepensis* reforestations in semi-arid Mediterranean environment. Furthermore, data from Spanish National Forest Inventory revealed that the medium cover of >40-years old *P. halepensis* plantations is 30 %. It is generally hypothesised that through vegetation growth an improvement of soil properties in reforested lands will develop.

However, results of analysis of soil properties in *Pinus* plantations are contrasting. While some studies suggest that planted forest may improve soil properties in decades, others have found lower soil organic content and nutrient concentration under *Pinus* plantations than in adjacent shrublands. It can be concluded that edaphologenetic processes are not fostered fast enough to counterbalance the initial impact of plantation techniques. The duality terrace/sidebank created a system where vegetation colonization of the most exposed microenvironment is severely hampered. In Southeast Spain, after 30 years the bank is hardly colonized, bare soil is 12 times more frequent than on the terrace. On the other hand, the species pool is severely depleted and species richness is less than half that on original slopes. The poor colonization may also be a consequence of the development of a recalcitrant litter layer. In fact, *Pinus halepensis* litter decomposes with difficulty, especially under the dry conditions of a semi-arid environment. This fact leads to a layer of litter that physically and allelopathically affects the germination of seeds and the growth of seedlings that may hamper demographic recovery of the understorey layer. New restorations should aim to delineate the maximum potential for vegetation in the landscape and develop appropriate vegetation strategies. Knowledge of species interactions and demographic processes have to be incorporated into the design of the restoration.

In Reforested lands, focus is on plants which can be planted to strengthen side banks and reduce runoff from terraces that contribute to headwater development of rills and gullies. The following three vegetation strategies are recommended to correct negative impacts and reduce connectivity in Reforested Lands.

1. Vegetation should be planted where the rills originate, where terraces are collapsing or across terraces not perpendicular to slope as these will form zones of runoff during rainfall events.
2. Side banks are much more extensive structures. Massive plantations of vegetation are expensive, and mostly unsuccessful because of the harsh conditions. Encouraging spontaneous or induced colonization should be the focus of the works. Microstructures built with natural barriers like wood, debris and stones should be established on the side banks in order to promote the trapping of water, nutrients and seeds. Conditions could be further improved by local addition of organic amendments (to improve soil infiltration and nutrient status) and seeding with side-bank adapted species.
3. In mature forests, some manipulation of litter layer and seeding of shrubs and grasses should encourage understorey development. This should be especially targeted at areas between trees that function as contributing headwater areas of rills and gullies.

In terms of species which should be considered, a combination of grasses and shrubs are recommended to provide the understorey cover in reforested lands. Of the grasses these are *Stipa tenacissima*, *Brachypodium retusum* and *Helictotrichon filifolium*. Shrub species recommended are *Salsola genistoides* on the side-banks, while for other spots *Rosmarinus officinalis* and *Anthyllis cytisoides* should be planted first, followed by *Rhamnus lycioides* and *Pistacia lentiscus*.

5.2.1.2   Rainfed Croplands

In rainfed croplands, the focus has been on assessing the feasibility of using cover crops and vegetation strips in preventing desertification of rainfed agriculture with water deficits. In developing criteria for selecting suitable cover crops, in addition to the general criteria for desirable plants (Table 5.1) the following points need to be considered:

1. Limited rooting depth (<= 30 cm) to reduce competition for water with perennial crops (Meerkerk et al. 2008).
2. Dominates undesired weeds
3. Annual species – perennial species are undesirable because in most cropping systems considered, the cover crops can only be grown during a part of the year.
4. Leguminous species because they may reduce fertilisation costs (Bowman et al. 2000; Ingels et al. 1998).

Olive orchards have one extra selection criterion for cover crops compared to almond orchards that is related to harvest operations. The olive oil harvest, which involves 97 % of the olive production, takes place from December to early March. In southern Europe it is still common to harvest by hand, spreading nets on the floor and beating the olives off the trees with sticks. This makes tall cover crops impractical. Therefore, the additional criterion is a limited height of the cover crop, ideally <= 10 cm. An alternative is to restrict the cover crop to the lanes in between the tree rows, or to remove the cover prior to harvesting.

Apart from the role of water availability, additional criteria must be met for the successful application of cover crops in rainfed agriculture. In semi-arid areas in general, the presence of cover crops on the field must be tuned to cultivation practices like sowing and harvesting performed on the main crop. In orchards and vineyards, cover crops can be grown throughout the year if there is enough water, whereas in annual cropping systems like winter cereals, the growth of cover crops will be restricted to the fallow year. Olive and almond plantations that are harvested by hand require a clean soil below the trees during the harvest period. Two specific issues can identified for the application of grassed waterways in thalwegs. The first is that thalwegs cross field borders and may require the cooperation of several landowners. In addition, grassed waterways can complicate tillage operations: more turning is required to leave the grass undisturbed.

Different types of cover crops can be considered, including weeds, legumes and grass species (Fig. 5.1). Ingels et al. (1998) have presented a selection of potential cover crops consisting mainly of winter annuals and legumes: in the target areas in Tuscany and Murcia, cover crops in summer are undesirable with regard to rainfed production. In addition, only species with a moderate to high seedling vigour should be selected.

The advantages and problems of use of cover crops in cropland are summarised in Table 5.2.

The selection and application of specific vegetation measures should be done on (sub)catchment level, taking into account the local climate, landscape and

Fig. 5.1 Potential cover crops include weeds (*left*), legumes (*middle*) and grasses (*right*)

Table 5.2 Major benefits and drawbacks of vegetation measures in cropland

	Benefits	Potential drawbacks
Cover crops/ vegetation strips	Effective protection against soil degradation and loss of soil productivity	Competition for water between cover crop and main crop
	Increased infiltration of rain water	Increase in production and equipment costs, especially when specific vegetation species are sown
	Improvement of soil structure	The green soil cover may lower the soil temperature in spring and increase the risk of frost damage in orchards
	The mulch left after chemical weeding prolongs the period of soil protection and decreases water loss by soil evaporation	
Grassed waterways	Reduced risk of gully formation	May require the cooperation of several landowners
	Reduction of runoff volume and peak discharge at (sub)catchment level	Reduced trafficability of fields

cropping systems. Figure 5.2 shows the priority areas to be protected during the rainy season: terraced orchards/vineyards when competition for water is high (left), orchards on steep slopes (middle) and sloping cereal fields (right). The dip slope of earth terraces can be stabilised by natural vegetation. This is a common and quite effective practice in many areas. The effect of cover crops on production will be limited when their growth is restricted to the thalwegs, even in dry areas.

Several water harvesting techniques can be used to increase the water availability for vegetation growth. For perennial crops, a method to maximise production is to adapt the crop spacing to the local climate (Tubeileh et al. 2004a). The roots of trees like almond and olive are able to mine the soil around their trunk. If the climate is drier and water availability lower, individual trees need a greater volume of soil to meet their water requirement. Hence, at sites with lower water availability, the trees need to be spaced further apart for optimal production. This explains the rationale behind traditional tree arrangements, with up to 24 m bare soil between rows of trees, as still present in Tunisia (Ennabli 1993). Another way to increase water availability and water use efficiency is to apply a mulch. The aim of the mulch is to reduce the soil evaporation.

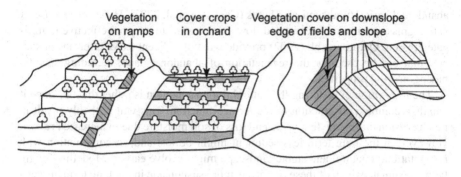

Fig. 5.2 Priority areas to be protected during the rainy season: terraced orchards/vineyards when competition for water is high (*left*), orchards on steep slopes (*middle*) and sloping cereal fields (*right*) (Hooke and Sandercock 2012)

5.2.1.3   Semi-Natural and Abandoned Lands

The *Criteria for selection of plant species* considered desirable for preventing terrace failure in semi-natural and abandoned lands are as identified in Table 5.1.

To mitigate soil erosion after agricultural land abandonment, and terrace failure specifically, the soil and water conservation practices can be divided into three groups: (1) maintenance of terraces and earth dams, (2) revegetation with indigenous species, and (3) restoring the original drainage pattern. The first option should include restoration of terrace walls after heavy rainfall and ploughing near the terrace wall to enhance the storage capacity and improve infiltration. This kind of management might even include subsidies for farmers who combine extensive agriculture with soil and water conservation. However, in the long term this option might not be very profitable, since the cost of subsidies will be high and when such a subsidy program stops the farmer might still abandon those fields, which means that erosion is only delayed.

For the second option of revegetation a distinction can be made between (i) revegetation on the terrace to improve infiltration and (ii) revegetation of terrace walls and zones with concentrated flow to prevent or mitigate gully erosion. Gyssels et al. (2005) conclude that for splash and sheet erosion vegetation cover is the most important parameter to control erosion, but for rill and gully erosion the effect of plant roots is at least as important. De Baets et al. (2006) demonstrate that especially grass roots are very effective in reducing soil detachment rates under concentrated flow. An increase in root density from 0 to 4 kg m^{-3} already decreases the relative soil detachment rates to very low values. Hence, vegetation species with a dense rooting system and good vegetation cover are most suitable for revegetation of terrace walls. Indigenous species that accomplish these characteristics are e.g. *Lygeum spartum*, *Brachypodium retusum* and *Stipa tenacissima*. However, other characteristics such as germination and growth rate are important as well for successful mitigation of terrace failure. From an ecological point of view the re-establishment of the indigenous shrub vegetation is a key step in the restoration of abandoned agricultural semi-arid lands (Caravaca et al. 2003). However, the use of native grass species in concentrated flow zones seems to be more effective to mitigate erosion. Quinton et al. (2002) provide a list of indigenous Mediterranean species that can be used for the revegetation of abandoned land, including specific bioengineering properties.

The last option of restoring the natural drainage pattern is rather drastic, since it entails a complete destruction of the terrace function. This will probably result in much erosion during the first years and loss of soil moisture due to increased runoff. However, in the long term less sediment might be lost and, in combination with revegetation, a natural and stable landscape might evolve earlier. Modelling might help to evaluate which of these options is more sustainable in the long term, as verification in the field is very problematic due to the diversity in topography, substrate and land use and the infrequent occurrence of extreme events. This makes it difficult to find field sites with similar conditions but with different land management. Modelling should elucidate which of the soil and water conservation practices will be most effective and sustainable in the long term in different locations, settings and environments.

5.2.2   Hillslopes and Gullies

To select species for gully erosion control, plant characteristics that influence the resistance to concentrated flow erosion must be selected. Traditionally, total plant cover and contact cover were used to evaluate species for controlling sheet and interrill erosion. However, for controlling concentrated flow erosion, new criteria have been investigated in more detail and quantified (Chap. 4) and are identified as:

Above-ground plant characteristics

- Stem density (SD, $m^2 m^{-2}$)
- Sediment trapping efficiency (TE, $m m^{-1}$)
- Resistance to flow shear forces (MEI, N)

Below-ground plant characteristics

- Root density (RD, $kg m^{-3}$) for the 0–0.1 m topsoil
- The fraction of fine roots (<5 mm) relative to total root mass (FR, %)
- Root area ratio (RAR, fraction) for the 0–0.1 m topsoil
- Root tensile strength (T_r, mPa) for 2 mm thick roots

These criteria address directly two erosion sub-processes:

- Hydraulic erosion
- Shallow mass movements occurring on gully banks

De Baets et al. (2006) demonstrated that grass roots are very effective in reducing soil detachment rates under concentrated flow. An increase in root density from 0 to 4 $kg m^{-3}$ already decreases the relative soil detachment rates to less than 5 % of

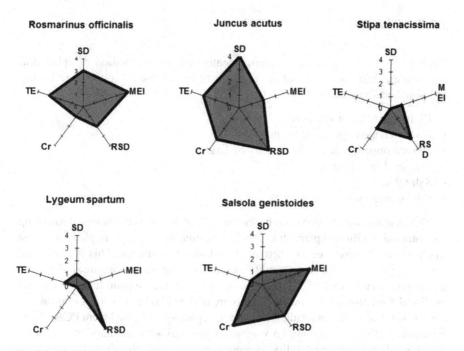

Fig. 5.3 Suitability of some Mediterranean plant species for rill and gully erosion control, based on their scores on the following criteria: Cr (kPa) is the root cohesion at 0.3–0.4 m soil depth, MEI (N) is the index of stiffness, SD (m² m⁻²) is the stem density, RSD (dimensionless) is the topsoil erosion reducing potential of plant roots during concentrated flow erosion and TE (m m⁻¹) is the trapping effectiveness (after De Baets et al. 2014)

the values for a rootless topsoil. Fine roots are found to be more effective in reducing concentrated flow erosion rates as compared to thick roots (De Baets et al. 2007a). The properties of some key plants suitable for use are shown in Fig. 5.3 in relation to the identified characteristics.

Planting species in rows perpendicular to the flow direction will provide an optimal resistance to erosion and a better ability to trap sediments and organic debris. Combining with the results of Rey (2003, 2004) it is recommended that on gully floors 50 % cover of low vegetation with high trapping effectiveness (e.g. *Juncus acutus*) in the gully floor renders the surface inactive (Rey 2003). Vegetation barriers can be established in gullies to further prevent soil loss; a downslope vegetation barrier covering only 20 % of this plot can be sufficient to trap all the sediments eroded upslope (Rey 2004) using species such as *Rosmarinus officinalis*. On terrace bank or gully walls planting species to prevent shallow mass movements should use deeply rooted shrubs such as *Anthyllis cytisoides, Salsola genistoides, Retama sphaerocarpa, Tamarix canariensis, Atriplex halimus* (De Baets et al. 2008).

5.2.3 Channels

Plant species were assessed on their potential use for restoration of ephemeral channels according to the list of desirable qualities in Table 5.1, plus the following additional characteristics:

- High threshold for removal
- Do not create large woody debris
- Not too obstructive/reduce capacity by large amount
- Perennial or persistent
- Salt tolerant
- Create even sward

The analysis of conditions and hydraulics (Chap. 4) forms the basis for assessing the potential of different plants for reducing erosion, the areas where plants are most likely to be effectively established and thresholds for removal. This is completed through a process of matching conditions for growth of plants with areas of erosion as demonstrated in Table 5.3. Suggested species within each plant functional group are listed. Conditions for growth of plants are outlined, as broken down into categories of 'substrate', 'water availability' and 'morphology' (derived from PCA). Field monitoring of impact of flows on vegetation, coupled with calculations of hydraulics is used to assess the 'ability to reduce erosion' and 'threshold for removal'. 'Areas of erosion' where plants may be established for the purpose of reducing erosion are indicated.

Of the grasses, *Lygeum spartum* is identified as having great potential for reducing erosion in channels. While it has been found to establish in a range of substrates, a preference for fines appears to exist. Extended monitoring indicates that this species is highly resistant to erosive floods, has the ability to trap sediments and contribute to the net aggradation of the channel bed. Within Cárcavo, these grasses are found in high density where there is a high supply of fine sediments along the channel network. This corresponds with two areas within the drainage basin, the first one in the upper part of the channel network in close proximity to gullies where there is a high supply of fine grained sediments; the second area is in the downstream zone immediately upstream of check dams where there are high rates of sedimentation. However, these latter areas also provide favourable conditions for other species, (ie Reeds and Tamarisk) that respond to greater water availability as water ponds behind these structures. It is suggested that *Lygeum spartum* is planted in close proximity to gullies contributing fines with the aim of trapping these incoming sediments.

As herbs have a very low resistance to erosion by channel flow, and are generally annuals, they are not considered to have potential for reducing erosion, although they may play a role in improving the organic matter content of the substrates which is beneficial for the establishment of other plants. Reeds, while they may be swept over by relatively minor flows, increase the resistance of the substrate to erosion and encourage sediment trapping. Their distribution is quite limited to low gradient

Table 5.3 Potential plants and identification of possible areas of erosion where they may be established in order to reduce erosion activity and sediment connectivity

Functional Group / Plant species	Qualities		Conditions for Growth			Ability to Reduce Erosion	Degradation Processes / Threshold for removal	Areas of erosion
	Desirable	Undesirable	Water Availability	Substrate	Morphology			
Grasses								
Lygeum spartum	High resistance to erosion / Ability to trap sediments / Can form dense stands / Perennial and persistent	Requires fine material	Perennial, tolerant to drought	Range of substrate, prefer fines	Thalweg, upper parts of channel network	Generally high, varies with substrate	Removal by flows>4 m.	Close proximity to incoming gullies contributing fines
Herbs								
Numerous species	Variable substrates	Annuals, not persistent	Annuals, not tolerant to drought	Range of substrates	Bed, bars and floodplain	Low	Removal by flows>1 m	–
Reeds								
Phragmites australis	High resistance to erosion	Require high moisture content	High water requirement, ponded areas	Range of substrates, prefer fines	Scoured areas, thalweg, low gradient	Low-Moderate	Swept over by flows<1 m	Low gradient reaches, thalweg in areas of ponding
Juncus sp	High thresholds	Invasive / Reduce biodiversity / Obstructive / Aesthetics / Fine sediment						

(continued)

Table 5.3 (continued)

Functional Group	Qualities		Conditions for Growth			Ability to Reduce Erosion	Degradation Processes	Areas of erosion
Plant species	Desirable	Undesirable	Water Availability	Substrate	Morphology		Threshold for removal	
Shrubs								
Dittrichia viscosa	Perennial; Coloniser on gravel bars; Ability to trap sediment in dense stands; Can form quite dense swards; Resprouts easily; Moderate resistance	Low ability to trap sediments	Drought tolerant	Gravels	Bar surfaces	Low	Removed by flows> 1 m	Gravel beds, bars and raised surfaces
Nerium oleander	High resistance to erosion; High thresholds for removal; High resistance to drought; Not invasive?; Perennial and persistent; Variable substrates; Sprouts easily but germination requires water	Highly poisonous; Ground cover usually limited; Some limited woody debris; Low ability to trap sediments	Drought tolerant	Gravels	Bar surfaces	Moderate	Removed by flows> 3 m	Gravel beds, bars and raised surfaces

Trees

Tamarix canariensis		Drought tolerant	Fines, areas of sediment storage	–	High	High resistance to removal, flows > 5 m.	Thalweg and bars, close proximity to check dams
Very high resistance to erosion	Creates large woody debris						
High thresholds for removal	Invasive						
	Obstructive						
	Prefers fine sediment						
Ability to trap sediments	May lead to excessive widening of channel, switching of flow, incision of non-Tamarisk part.						
Perennial and persistent							
Withstand drought							
Can create even cover but high level canopy							
Germinates and propagates easily							
Salt tolerant							

reaches, where ponding occurs, this providing the extra water required for their survival. Experimental studies indicate that *Phragmites australis* has a marked effect in reducing erosion, with high stem densities increasing the retention of sediments and the finely distributed mat system protecting the surface against erosion (Coops et al. 1996). Even when swept over by floods, the recovery of reed communities can be rapid, with above ground biomass being replaced very quickly through the emergence and growth of new shoots. Establishment of reeds in low gradient reaches and thalweg where there are freshwater ponds may be effective in increasing the resistance of the bed to erosion and trapping sediments.

The two common shrub species that could have potential for erosion reduction, *Dittrichia viscosa* and *Nerium oleander* both have a preference to gravel substrates. However, *Dittrichia viscosa* has quite a low resistance to erosion and is generally removed by moderate sized floods (>1 m). Its ability to trap sediments is also quite limited due to its size and form. *Nerium oleander* can form a significant roughness element in the channel, serving to reduce the velocity of flows through a reach. Whilst its ability to trap sediments is considered quite low, its resistance to erosion is rated high. A major flood (>3 m) would be required to sweep over a mature individual, and, unless it is scoured out at its base through the removal of surrounding substrate, it is likely to recover. Individual bushes also add considerable roughness value to the channel and are therefore effective in reducing the velocity and power of floods along these channels. It is suggested that these be encouraged to grow in areas of sediment transport and storage (gravel beds, bars and raised surfaces).

Tamarix canariensis is by far the dominant native tree species present along these channels; its distribution has increased as a result of the construction of check dams. The large volumes of fine grained sediments stored upstream of the structures, and ponding of water that occurs in these zones provides favourable conditions for plant establishment. They also have high resistance to long periods of drought and high salinity levels, and are therefore well suited to the extremes of conditions that characterise these channels (DiTomaso 1998). Their ability to reduce erosion and resistance to removal is rated as high. Areas of erosion where they may planted include the thalweg and bars, and in close proximity to check dams. However, some concern exists about its use as they are highly invasive, can dominate the riparian community and in larger channels they have been associated with increased braiding and avulsion in south-western USA (Graf 1983). Further consideration is required on whether similar problems could be expected to arise within the ephemeral channels in Southeast Spain. The aesthetic quality has also been questioned. *Retama sphaerocarpa* has similar properties to Tamarisk and grows predominantly in schistose substrate channels of Southeast Spain.

The threshold for removal appears to be very high, though they can be removed by lateral incision, but if they are removed then they have the potential to form large woody debris, now considered beneficial on some channels, but long considered a flood hazard in some regions (e.g. Italy). Even if *Tamarix canariensis,* and similarly *Retama sphaerocarpa*, are severely damaged by high flows they have a strong capacity to resprout from damaged pieces and this may relatively quickly revegetate an area.

Within the channels, we have identified the following as goals to guide planting strategies: 1.minimise the delivery of sediments to the channel and maximise sedimentation within channel, but not to the extent that it causes problems (ie. erosion downstream by clear water flows); 2. minimise erosion along the channel through the effect vegetation has in reducing velocity of flows and increasing overall resistance of channel to erosion.

The following areas are identified as erosion hotspots in channels: incoming gullies; confluences; thalweg; areas downstream from check dams; valley walls undercut by the stream. Vegetation can potentially mitigate erosion and reduce sediment connectivity in all these sites with the exception of active valley walls, which are generally too steep. Valley walls represent a significant source of sediments. Channel widening through erosion of valley side walls is the most effective means a channel has to reduce its force. If this was to be restricted, then the channel will be forced to incise its bed at an accelerated rate. It is critical that planting strategies are based on a sound understanding of the local environmental conditions, particularly substrate and water availability and existing sediment connectivity, but also position within the channel and exposure to high forces. The likely influence that planted vegetation (and any associated measures) will have on future connectivity and possible sediment deficit also needs to be considered.

While the aim is also to minimise the amount of structural interventions, it may be advantageous to have some structural measures which will favour vegetation establishment. For example, construction of a small log/stone structure across incoming gully/thalweg of channel and planting of vegetation immediately upstream of the structure in an area where sedimentation is expected to occur could be effective. This may be necessary in the situation where there is insufficient substrate at a site for planting (i.e. site eroded to bedrock, high connectivity of sediments through reach). Sedimentation induced upstream by the structure and change(s) in moisture status may alter the conditions such as to improve seedling establishment. Such bioengineering measures have been trialled in marly gullies in the southern Alps, France, the improvement in soil water conditions increasing the capacity of the sediment deposits to allow germination (Rey 2006).

The following four areas are highlighted as areas where vegetation may be encouraged to mitigate erosion and reduce sediment connectivity. Possible bioengineering measures are proposed.

1. Incoming gullies, where the gully has incised to bedrock and is bare. Construction of small log/stone structure across floor of gully and planting of vegetation in sedimentation area upstream. Choice of planting depends on substrate, but if fines: *Lygeum spartum* and *Tamarix canariensis*.
2. Fan from gully/tributary, sediments are input to the channel and deposited immediately downstream from the confluence as a fan. Construction of stone bund around the edge of the fan and planting of vegetation in sedimentation area upstream. Choice of planting depends on substrate. If coarse gravels: *Dittrichia viscosa* and *Nerium oleander*, if fines: *Lygeum spartum* and *Tamarix canariensis*.

3. Small thalweg of channel. Construction of logs/stone structure across thalweg and planting of vegetation in sedimentation area upstream. Choice of planting depends on substrate, but if fines: *Lygeum spartum* and *Tamarix canariensis*.

4. Areas downstream of check dams. Very difficult to establish vegetation in these areas if there is no sediment. Suggest that vegetation is also planted in areas immediately downstream of check dam at time of construction. If coarse gravels: *Dittrichia viscosa* and *Nerium oleander*, if fines: *Lygeum spartum* and *Tamarix canariensis*. If it is an area where water ponding occurs, consider planting reeds such as *Phragmites australis* and *Juncus* species.

5.2.4   Catchment Scale and Synthesis

These various strategies can be put together at the small catchment scale in a spatially integrated strategy. An example of this is given for a small sub-catchment area of Cárcavo, SE Spain (Fig. 5.4), which includes field terraces and a track crossing the catchment. The area is mostly a mixture of almond and olive trees with a small area of abandoned land. Pathways of runoff and erosion occurring in several events (red arrows) and areas where sediments were deposited (grey areas) are shown. The steep uplands and area of abandoned land were a significant source of runoff. Many of the larger rills commenced immediately east of the road, with connecting pathways following the natural drainage line that exists in the landscape. There was also significant runoff along the elongated terraces as they are not constructed on the contour.

Figure 5.4 outlines suggested strategies that could be applied to the hotspot areas and pathways to reduce the potential connectivity in the landscape. These can be summarised as follows:

• Establish more vegetation on terrace banks. Suitable species include grasses like *Lygeum spartum, Brachypodium retusum* and *Stipa tenacissima* in combination with more deeper rooted shubs like *Anthyllis cytisoides, Atriplex halimus* or *Salsola genistoides*.

• Plant vegetation on flat terraces of abandoned lands and vegetate banks. Suitable species include *Lygeum spartum, Brachypodium retusum* and *Stipa tenacissima*.

• Plant cover crops in tree lanes of all fields in winter time (cover up to 50% of field area). Planted cover crops should follow the contour as much as possible. Planting winter annuals and weeds is recommended.

• Plant vegetation at sides of tracks. Suitable species include *Brachypodium retusum*.

• Plant vegetation along the natural drainage line (grassed waterway). Suitable grass species include *Brachypodium retusum*. Where water accumulates plant *Juncus sp*.

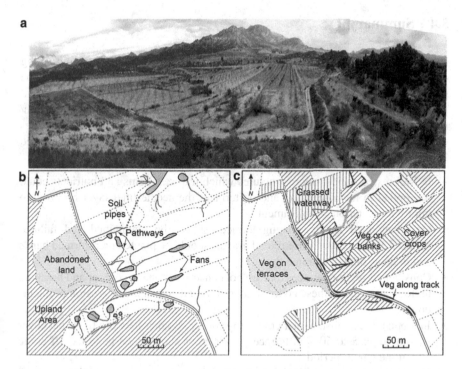

Fig. 5.4 (**a**) Photograph of a sub-catchment of Cárcavo, (**b**) Mapped connectivity pathways, (**c**) Suggested strategy for erosion reduction and increase of sedimentation by connectivity minimisation (after Hooke and Sandercock 2012)

5.3 Guidelines

A set of guidelines has been compiled based on detailed studies of vegetation and its positive effects in mitigating erosion and reducing connectivity at a range of scales. The purpose of these guidelines is to provide information on how problems of soil erosion and land degradation may be controlled by the application of innovative vegetation strategies with existing soil conservation measures. These strategies target specifically those hotspot areas in the landscape where erosion is a problem at present, or which, if improperly managed, will become a significant problem. The approach is different from other approaches in that it identifies hotspots and focuses on the application of appropriate vegetation species to these areas and their position in the landscape, whereas other approaches are applied across the entire landscape. These guidelines are suitable for dryland environments of the Mediterranean region of southern Europe, and they are based on research in Southeast Spain. These guidelines are presented in the Supplementary Material.

5.4   Summary

In this chapter a range of vegetation strategies has been suggested for mitigating erosion and reducing connectivity, and these are also communicated within the guidelines document which was disseminated to landowners, land managers and policy makers In Spain and Italy and is available at http://www.port.ac.uk/research/recondes/practicalguidelines/. These strategies can be summarised below as a series of recommendations, which apply to particular land units and hotspots, as listed below.

- In Reforested Lands, microstructures which trap sediments and nutrients could be applied to the side banks and on terraces the needle litter layer removed to improve vegetation establishment.
- On Abandoned Lands existing terraces should be maintained and failures repaired after events. Dense rooted grasses should be planted on reformed banks and in areas of concentrated flow.
- Consider correcting terraces that are not constructed on the contour and breaking up long terraces, as these concentrate runoff and contribute to erosion problems downslope.
- In croplands establish winter cover crops in access lanes between trees, but cover should not exceed 50 % of terrace. Cover crops should then be killed off at the end of the winter period.
- Plant vegetation cover along edges of tracks, particularly where these cross a drainage line or gradient changes such that the track begins to concentrate runoff.
- Where there are natural drainage lines crossing fields establish grassed water-ways. Double drilling techniques should also be used when seeding these areas. Establish vegetation on gully floors.
- In small channels, revegetation efforts should focus on planting grasses in areas where there are fine sediment inputs. Small structures may be built to promote deposition and improve conditions for vegetation establishment.
- In larger channels, efforts should focus on establishing larger shrubs and trees as these will have a greater effect in reducing flow velocities and trapping sediments, therefore reducing sediment connectivity to areas downstream.

5.5   Wider Application and Global Implications

This approach of minimising connectivity and thus reducing soil erosion and sediment movement using vegetation in strategic spatial locations has been developed for the Mediterranean environment but is much more widely applicable. It is applicable to lands where the dominant degradation is by water. In such areas the patterns and locations of flow pathways and erosion hotspots can be identified and the contributors to connectivity targeted. Many traditional farming systems have recognised the need to minimise lengths and gradients of slope and have thus used terracing or similar techniques. However, increased mechanisation and need for efficiency has meant many of these structures and practices are being abandoned or removed. Similar principles of identifying patterns of flow lines and zones of erosion apply in non-terraced landscapes.

The strategies and recommendations have been developed in relation to marl bedrock areas. Areas of this soft rock are particularly vulnerable to soil erosion and land degradation. However, the approach would be applicable to other soft rock areas and zones with vulnerable soils. Local research would need to be undertaken to identify suitable local plant species and test their properties. In some areas the soil and soft rock is susceptible to piping, especially in clay soil, and this can be a major contributor to land degradation, as in many parts of southern Spain. The piping is frequently associated with rills and with drainage lines, and with terraces on abandoned land. Indeed, in the area illustrated in Fig. 5.4. there was some large piping in the lower part of the hillslope (Hooke 2006). The connectivity approach can be applied to understand the pathways (Marchamalo et al. 2016) and these can then be targeted.

Likewise, these ideas and approach could be transferred to other regions of the world but it would require investigation or application of knowledge of local plants and their suitability for various locations to ensure optimal effectiveness and to avoid use of invasive, exotic or unsuitable plants. The strategy has been developed in relation to agricultural areas but it may be applicable to grazed areas. However, this is challenging because of the consumption of plants by the animals. In some areas, e.g. Ethiopia, exclosures have been successful. A key advantage of the strategy advocated here is that it does not take up large areas of land and make them unproductive. The key locations for planting are on margins of fields, on structures, tracks etc. Patterns of animal movement may also need to be considered on grazing lands but their relation to drainage and runoff lines can be examined. Strategic fencing of key areas could still be used in grazing lands.

In terms of climatic regions, those most vulnerable to increased desertification and land degradation are the arid margins of deserts, semi-arid lands and to some extent sub-humid lands. As long as the native plants and the processes are examined then the approach can be used. It is not applicable in areas prone to wind erosion and aeolian land degradation. There are other locations in which vegetation is removed, leading to land degradation and there is a need for land restoration using vegetation, including contaminated and industrial land but the latter have been subject to much research. The approach and recommendations here have not been developed in relation to intensively irrigated land and extensive lowlands where salinisation may the major problem and cause of land degradation. Application in such environments needs further consideration of water balances, salt tolerant species and groundwater relations. If surface water flow is not the problem then the connectivity approach advocated here is less applicable.

Scenarios of both climate change and land use change and their combination all point towards likely increased desertification and land degradation in many parts of the world. This is at a time of increased population and concern about future food security so land degradation is likely to increase without such measures as recommended here being taken.

Any implementation of the approach and strategies suggested here requires the cooperation of the land users and owners. The ideas need to be introduced and discussed within communities and adopted by them. It is suggested that local people could be readily trained up to identify signs of flow lines and erosion and that simple systems of mapping using facilities such as GoogleEarth could be implemented in a wide range of environments and communities. The approach is dependent on identifying the connectivity and thus whole patterns in the landscape. Any scheme will entail cooperation between farmers in areas so that the whole connected system is considered. The consequences of any disconnection, especially on water supply, need to be considered. However, such an approach is likely to have multiple benefits because, if water infiltrates to a greater extent, it is likely to sustain local water supplies more effectively and is likely to reduce risk of flooding from excessive overland flow.

The key to application of these spatial strategies is adaptation to the particular landscape. This entails understanding of the processes and their spatial patterns and matching attributes of the soil and plants to these in optimal locations for land degradation reduction.

References

Aldridge BN, Garrett JM (1973) Roughness coefficients for stream channels in Arizona: U.S. Geological Survey Open-File Report. p 87

Allen RG, Pereira LS, Raes D, Smith M (1998) Crop evapotranspiration – guidelines for computing crop water requirements. FAO, Rome

AMINAL (2002) Werk maken van Erosiebestrijding. Ministerie van de Vlaamse Gemeenschap, administratie Milieu- Natuur- Land- en Waterbeheer, afdeling Land Brussel

Aragüés R, Puy J, Isidoro D (2004) Vegetative growth response of young olive trees (Olea europaea L. cv. Arbequina) to soil salinity and waterlogging. Plant Soil 258:69–80

Arcement GJ, Schneider VR (1989) Guide to selecting Manning's roughness coefficients for natural channels and floodplains, United States geological survey water supply paper, 2339., pp 1–38

Baker WL (1989) Macro- and micro-scale influences on riparian vegetation in western Colorado. Ann Assoc Am Geogr 79:65–78

Barberá GG, López Bermúdez F, Romero Díaz A (1997) Cambios de uso del suelo y desertificación: el caso del sureste ibérico. In: García Ruiz JM, López García P (eds) Acción humana y desertificación en ambientes mediterráneo. Instituto Pirenaico de Ecología, Zaragoza, pp 9–39

Baskin CC, Baskin JM (1998) Seeds: ecology, biogeography, and evolution of dormancy and germination. Academic Press, San Diego, p 676

Bastida F, Barberá GG, García C, Hernández T (2008) Influence of orientation, vegetation and season on soil microbial and biochemical characteristics under semiarid conditions. Appl Soil Ecol 38(1):62–70

Beaufoy G (2002) The environmental impact of olive oil production in the European Union: practical options for improving the environmental impact. European Forum on Nature Conservation and Pastoralism

Bellin N (2006) Impacts de la modification du parcellaire sur la connectivité hydrologique et l'érosion hydrique d'un bassin versant situé dans un milieu semi-aride (région de Murcie, Espagne)

Bellin N, van Wesemael B, Meerkerk A, Vanacker V, González Barberá G (2009) Abandonment of soil and water conservation structures in Mediterranean ecosystems. A case study from south east Spain. Catena 76:114–121

Bendix J, Hupp CR (2000) Hydrological and geomorphological impacts on riparian plant communities. Hydrol Process 14:2977–2990

Birkeland GH (1996) Riparian vegetation and sandbar morphology along the lower little Colorado river, Arizona. Phys Geogr 17(6):534–553

© Springer International Publishing Switzerland 2017
J. Hooke, P. Sandercock, *Combating Desertification and Land Degradation*,
SpringerBriefs in Environmental Science, DOI 10.1007/978-3-319-44451-2

Bischetti GB, Chiaradia EA, Simonato T, Speziali B, Vitali B, Vullo P, Zocco A (2005a) Root strength and root area ratio of forest species in Lombardy (Northern Italy). Plant Soil 278:11–22

Bischetti GB, Chiaradia EA, Simonato T, Speziali B, Vitali B, Vullo P, Zocco A (2005b) Root strength and root area ratio of forest species in Lombardy (Northern Italy). Plant Soil 278:11–22

Bochet E, Garcia-Fayos P, Tormo J (2007) Road slope revegetation in semiarid mediterranean environments. Part I: Seed dispersal and spontaneous colonization. Restor Ecol 15(1):88–96

Boix-Fayos C, Barberá GG, López-Bermúdez F, Castillo VM (2007) Effects of check dams, reforestation and land-use changes on river channel morphology: case study of the Rogativa catchment (Murcia, Spain). Geomorphology 91:103–123

Boix-Fayos C, de Vente J, Martínez-Mena M, Barberá GG, Castillo V (2008) The impact of land use change and check-dams on catchment sediment yield. Hydrol Process 22:4922–4935

Boix-Fayos C, Martínez-Mena M, Pérez-Cutillas P, de Vente J, Barberá GG, Mosch W, Navarro-Cano J, Navas A (in press) Carbon redistribution by erosion processes in a intensively disturbed catchment. Catena

Bonet A (2004) Secondary succession of semi-arid Mediterranean old-fields in south-eastern Spain: insights for conservation and restoration of degraded lands. J Arid Environ 56:213–233

Borselli L, Pellegrini S, Torri D, Bazzoffi P (2002) Tillage erosion and land levelling: evidences in Tuscany (Italy). In: Rubio JL, Morgan RPC, Asins S, Andreu V (eds) Man and soil at the third millenium. Geoforma Ediciones – CIDE, Logroño, pp 1341–1350

Borselli L, Torri D, Oygarden L, De Alba S, Martínez-Casasnovas JA, Bazzoffi P, Jakab G (2006) Land levelling. In: Boardman J, Poesen J (eds) Soil erosion in Europe. Wiley, Chichester, pp 643–658

Borselli L, Cassi P, Torri D (2008) Prolegomena to sediment and flow connectivity in the landscape: A GIS and field numerical assessment. Catena 75(3):268–277

Borselli L, Salvador Sanchism P, Batolini D, Cassi P, Lollino P (2011) PESERA-L model: an addendum to the PESERA model for sediment yield due to shallow mass movement in a watersheed: CNR-IRPI, Italy Report.n.82. scientific report deliverable 5.2.1 DESIRE. Project. p 28

Bowman G, Shirley C, Cramer C (2000) Managing cover crops profitably. Sustainable Agriculture Network, Beltsville

Bracken LJ, Croke J (2007) The concept of hydrological connectivity and its contribution to understanding runoff-dominated geomorphic systems. Hydrol Process 21(13):1749–1763. doi:10.1002/hyp.6313

Bracken LJ, Wainwright J, Ali GA, Tetzlaff D, Smith MW, Reaney SM, Roy AG (2013) Concepts of hydrological connectivity: research approaches, pathways and future agendas. Earth Sci Rev 119:17–34

Brierley G, Fryirs K (1998) A fluvial sediment budget for upper Wolumla Creek, South Coast, New South Wales, Australia. Aust Geogr 29(2):107–124

Brierley G, Stankoviansky M (2002) Geomorphic responses to land use change: lessons from different landscape settings. Earth Surf Process Landf 27:339–341

Brierley G, Fryirs K, Jain V (2006) Landscape connectivity: the geographic basis of geomorphic applications. Area 38(2):165–174

Brock JH (1994) Tamarix (salt cedar), an invasive exotic woody plant in arid and semi-arid riparian habitats of western U.S.A. In: Waal LC, Child LE, Wade PM, Brock JH (eds) Ecology and management of invasive riverside plants. Wiley, Chichester, pp 27–44

Brown CB (1944) Sedimentation in reservoirs. Trans Am Soc Civ Eng 109:1047–1106

Burylo M, Rey F, Delcros P (2007) Abiotic and biotic factors influencing the early stages of vegetation colonization in restored marly gullies (Southern Alps, France). Ecol Eng 30(3):231–239

Cammeraat LH (2002) A review of two strongly contrasting geomorphological systems within the context of scale. Earth Surf Process Landf 27(11):1201–1222

Cammeraat LH (2004) Scale dependent thresholds in hydrological and erosion response of a semi-arid catchment in southeast Spain. Agric Ecosyst Environ 104:317–332

Cammeraat LH, Imeson AC (1999) The evolution and significance of soil-vegetation patterns following land abandonment and fire in Spain. Catena 37:107–127

Cammeraat LH, van Beek LPH, Dorren L, Torri D, de Baets S, Poesen J, Hooke J, Sandercock S, Stokes A, Sharman M, Norris J, Greenwood J, Andreu V, Clark S, Imeson A, van Wesemael B, J., B (2005) RECONDES field protocol version 1.0: unpublished Recondes project report, University of Amsterdam, p 58

Cammeraat LH, van Beek LPH, Dooms T (2009) Modelling water and sediment connectivity patterns in a semi-arid landscape. In: Romero-Diaz A, Belmont-Serrano F, Alonso-Sarria F, Lopez-Bermudez F (eds) Advances in studies on desertification. Edit Um, Murcia, pp 105–108

Cammeraat ELH, Cerdà A, Imeson AC (2010) Ecohydrological adaptation of soils following land abandonment in a semi-arid environment. Ecohydrology 3:421–430

Caravaca F, Alguacil MM, Figueroa D, Barea JM, Roldan A (2003) Re-establishment of Retama sphaerocarpa as a target species for reclamation of soil physical and biological properties in a semi-arid Mediterranean area. For Ecol Manag 182:49–58

Carson WP, Peterson CJ (1990) The role of litter in an old-field community: impact of litter quantity in different seasons on plant sciences richness and abundance. Oecologia 85:8–13

Castillo V, Barberá GG, Mosch W, Navarro-Cano JA, Conesa C, López-Bermúdez F (2001) Monitoring and evaluation of hydrologic-forestal restoration projects. In: López-Bermúdez F (ed) Monitoring and evaluation of the effects on environment of drought and erosion processes in the Region of Murcia. Consejería de Agricultura, Agua y Medio Ambiente de la Región de Murcia, Murcia

Castillo VM, Mosch WM, Conesa Garcia C, Barberá GG, Navarro Cano JA, López-Bermúdez F (2007) Effectiveness and geomorphological impacts of check dams for soil erosion control in a semiarid Mediteranean catchment: El Cárcavo (Murcia, Spain). Catena 70(3):416–427

Ceballos A, Schnabel S (1998) Hydrological behaviour of a small catchment in the dehesa landuse system (Extremadura. SW Spain). J Hydrol 210:146–160

Cerdà A (1997) Soil erosion after land abandonment in a semiarid environment of Southeastern Spain. Arid Soil Res Rehabil 11:163–176

Cerdà A (2007) Soil water erosion on road embankments in Eastern Spain. Sci Total Environ 378:151–155

Cerdà A, Giménez-Morera A, Bodí MB (2009) Soil and water losses from new citrus orchards growing on sloped soils in the Western Mediterranean basin. Earth Surf Process Landf 34:1822–1830

Cerdà A, Hooke J, Romero-Diaz A, Montanarella L, Lavee H (2010) Preface: soil erosion on Mediterranean type-ecosystems. Land Degrad Dev 21(2):71–74

Chaparro J, Esteve MA (1995) Evolución geomorfológica de laderas repobladas mediante aterrazamientos en ambientes semiáridos (Murcia, SE de España). Cuaternario y Geomorfología 9:39–49

Clarke ML, Rendell HM (2000) The impact of the farming practice of remodelling hillslope topography on badland morphology and soil erosion processes. Catena 40:229–250

Coops H, Geilen N, Verheij HJ, Boeters R, van der Velde G (1996) Interactions between waves, bank erosion and emergent vegetation: an experimental study in a wave tank. Aquat Bot 53(3–4):187–198

Craig MR, Malanson GP (1993) River flow events and vegetation colonization of point bars in Iowa. Phys Geogr 14:436–448

Croke J, Mockler S, Fogarty P, Takken I (2005) Sediment concentration changes in runoff pathways from a forest road network and the resultant spatial pattern of catchment connectivity. Geomorphology 68(3–4):257–268

D'Antonio C, Meyerson LA (2002) Exotic plant species as problems and solutions in ecological restoration: a synthesis. Restor Ecol 10(4):703–713

De Baets S, Poesen J (2010) Empirical models for predicting the erosion-reducing effects of plant roots during concentrated flow. Geomorphology 118:425–432

De Baets S, Poeson J, Gyssels G, Knapen A (2006) Effects of grass roots on the erodibility of topsoils during concentrated flow. Geomorphology 76:54–67

De Baets S, Poesen J, Knapen A, Galindo P (2007a) Impact of root architecture on the erosion-reducing potential of roots during concentrated flow. Earth Surf Process Landf 32(9):1324–1345

De Baets S, Poesen J, Knapen A, Gonzáles Barberá G, Navarro JA (2007b) Root characteristics of representative Mediterranean plant species and their erosion-reducing potential during concentrated runoff. Plant Soil 294:169–183

De Baets S, Poesen J, Reubens B, Wemans K, De Baerdemaeker J, Muys B (2008) Root tensile strength and root distribution of typical Mediterranean plant species and their contribution to soil shear strength. Plant Soil 305:207–226

De Baets S, Poesen J, Reubens B, Muys B, De Baerdemaeker J (2009) Methodological framework to select plant species for controlling rill and gully erosion. Earth Surf Process Landf 34:1374–1392

De Baets S, Quine TA, Poesen J (2014) Root strategies for rill and gully erosion control. In: Morte A, Varma A (eds) Root engineering: basic and applied concepts, vol 40, Soil biology. Springer-Verlag, Berlin/Heidelberg, p 493

De Graaff J, Eppink LAAJ (1999) Olive oil production and soil conservation in Southern Spain, in relation to EU subsidies. Land Use Policy 16:259–267

Di Tomaso JM (1998) Impact, biology, and ecology of saltcedar (Tamarix spp.) in the Southwestern United States. Weed Technol 12(2):326–336

Díaz-Ambrona CH, Mínguez IM (2001) Cereal-legume rotations in a Mediterranean environment: biomass and yield production

Dunjo G, Pardini G, Gispert M (2003) Land use change effects on abandoned terraced soils in a Mediterranean catchment, NE Spain. Catena 52:23–37

Ennabli N (1993) Les aménagements hydrauliques et hydro-agricoles en Tunisie: Institut national agronomique de Tunis. Département du génie rural des eaux et des forêts, Tunis

Eshel A, Hening-Sever N, Ne'eman G (2000) Spatial variation of seedling distribution in an east Mediterranean pine woodland at the beginning of post-fire succession. Plant Ecol 148:175–182

Evrard O, Bielders CL, Vandaele K, van Wesemael B (2007) Spatial and temporal variation of muddy floods in central Belgium, off-site impacts and potential control measures. Catena 70(3):443–454. doi:10.1016/j.catena.2006.11.011

Facelli JM (1994) Multiple indirect effects of plant litter affect the establishment of woody seedlings in old fields. Ecology 75(6):1727–1735

Facelli JM, Pickett STA (1991) Plant litter: light interception and effects an old-field plant community. Ecology 72(3):1024–1031

Ferguson L, Sibbett G, Martin G (1994) Olive production manual. University of California, Division of Agriculture and Natural Resources, Oakland

Fryirs K, Brierley G (2000) A geomorphic approach to the identification of river recovery potential. Phys Geogr 21(3):244–277

Fryirs KA, Brierley GJ, Preston NJ, Kasai M (2007a) Buffers, barriers and blankets: the (dis)connectivity of catchment-scale sediment cascades. Catena (Giessen) 70:49–67

Fryirs KA, Brierley GJ, Preston NJ, Spencer J (2007b) Catchment-scale (dis)connectivity in sediment flux in the upper Hunter catchment, New South Wales, Australia. Geomorphology 84:297–316

Gálvez M, Parra MA, Navarro C (2004) Relating tree vigour to the soil and landscape characteristics of an olive orchard in a marly area of southern Spain. Sci Hortic 101:291–303

Garcia-Fuentes A, Salazar C, Torres JA, Cano E, Valle F (2001) Review of communities of Lygeum spartum L. in the south-eastern Iberian peninsula (western Mediterranean). J Arid Environ 48(3):323–339

García-Orenes F, Cerdà A, Mataix-Solera J, Guerrero C, Bodí MB, Arcenegui V, Zornoza R, Sempere JG (2009) Effects of agricultural management on surface soil properties and soil-water losses in eastern Spain. Soil Tillage Res 106(1):117–123

Gasque M (1999) Colonización del esparto (Stipa tenacissima L.) en zonas degradadas de clima semiárido. Ph.D., Polytechnic University of Valencia

Giménez Morera A, Ruiz Sinoga JD, Cerdà A (2010) The impact of cotton geotextiles on soil and water losses in Mediterranean rainfed agricultural land. Land Degrad Dev 21(2):210–217

Gómez-Plaza A, Martinez-Mena M, Albaladejo J, Castillo A (2001) Factors regulating spatial temporal distribution of soil water content in small semiarid catchments. J Hydrol 253:211–226

González-Hidalgo JC, De Luis M, Raventós J, Sánchez JR (2001) Spatial distribution of seasonal rainfall trends in a western Mediterranean area. Int J Climatol 21:843–860

Graf WL (1983) Flood-related channel change in an arid-region river. Earth Surf Process Landf 8:125–139

Gyssels G, Poesen J (2003) The importance of plant root characteristics in controlling concentrated erosion rates. Earth Surf Process Landf 28:371–384

Gyssels G, Poesen J, Bochet E, Li Y (2005) Impact of plant roots on the resisistance of soils to erosion by water: a review. Prog Phys Geogr 29(2):1–28

Hara T, Van der Toorn J, Hook JH (1993) Growth dynamics and size structure of shoots of *Phragmites australis*, a clonal plant. J Ecol 81:47–60

Hardy T, Panja P, Mathias D (2005) WINXSPRO, a channel cross-section analyzer, user's manual, version 3.0, General Technical Report RMRS-GTR-147. U.S. Department of Agriculture, Forest Service, Rocky Mountain Research Station, Fort Collins, p 94

Harper JL (1977) Population biology of plants. Academic Press, London

Harris RR (1987) Occurrence of vegetation on geomorphic surfaces in the active floodplain of a Californian stream. Am Midl Nat 118:393–405

Herrera J (1991) The reproductive-biology of a riparian Mediterranean shrub, Nerium-Oleander L (Apocynaceae). Bot J Linn Soc 106(2):147–172

Hooke J (2003) Coarse sediment connectivity in river channel systems; a conceptual framework and methodology. Geomorphology 56(1–2):79–94

Hooke J (2006) Human impacts on fluvial systems in the Mediterranean region. Geomorphology 79:311–335

Hooke JM (2007a) Monitoring morphological and vegetation changes and flow events in dryland river channels. Environ Monit Assess 127(1):445–457

Hooke JM (ed) (2007b) Conditions for restoration and mitigation of desertified areas using vegetation (RECONDES): review of literature and present knowledge. Office for Official Publications of the European Commission, Luxembourg, p 297

Hooke JM (2015) Variations in flood magnitude-effect relations and the implications for flood risk assessment and river management. Geomorphology 251:91–107

Hooke JM, Mant JM (2000) Geomorphological impacts of a flood event on ephemeral channels in SE Spain. Geomorphology 34(3–4):163–180

Hooke J, Mant JM (2001) Floodwater use and management strategies in valleys of Southeast Spain. Land Degrad Dev 13:165–175

Hooke J, Mant J (2002) Morpho-dynamics of ephemeral streams. In: Bull LJ, Kirkby MJ (eds) Dryland rivers: hydrology and geomorphology of semi-arid channels. Wiley, Chichester, pp 173–204

Hooke J, Mant J (2015) Morphological and vegetation variations in response to flow events in rambla channels of SE Spain. In: Dykes AP, Mulligan M, Wainwright J (eds) Monitoring and modelling dynamic environments (A Festschrift in memory of Professor John B. Thornes). Wiley, Chichester, pp 61–98

Hooke J, Sandercock P (2012) Use of vegetation to combat desertification and land degradation: recommendations and guidelines for spatial strategies in Mediterranean lands. Landsc Urban Plan 107:389–400

Hupp CR, Osterkamp WR (1985) Bottomland vegetation distribution along Passage Creek, Virginia, in relation to fluvial landforms. Ecology 66:670–681

Imeson AC, Prinsen HAM (2004) Vegetation patterns as biological indicators for identifying runoff and sediment source areas for semi-arid landscapes in Spain. Agric Ecosyst Environ 104:333–342

Ingels CA, Bugg RL, McGourty GT, Christensen LP (1998) Cover cropping in vineyards – a growers handbook. University of California, Division of Agriculture and Natural Resources, Oakland

IPCC (2007) Climate change 2007: The physical science basis; contribution of Working Group I to the Fourth Assessment Report of the Intergovernmental Panel on Climate Change. Cambridge University Press, New York

Izhaki I, Hening-Sever N, Ne'eman G (2000) Soil seed banks in Mediterranean Aleppo pine forests: the effect of heat, cover and ash on seedling emergence. J Ecol 88:667–675

Joffre R, Vacher J, Delosllanos C, Long G (1988) The dehesa – an agrosilvopastoral system of the Mediterranean region with special reference to the Sierra Morena area of Spain. Agrofor Syst 6(1):71–96

Johnson RR, Simpson JM (1985) Desertification of wet riparian ecosystems in arid regions of the North American Southwest. Paper presented at the Arid Lands-Today and Tomorrow. Proceedings of an International Research and Development Conference, 20–25 October 1985.

Kitajima K, Fenner M (2000) Ecology of seedling regeneration. In Fenner M(ed) Seeds: the ecology of regeneration in Plant Communities, 2nd edn, Wallingford, Commonwealth Agricultural Bureau International, pp 331–360. 397 pp

Kosmas C, Gerontidis S, Marathianou M (2000) The effect of land use change on soils and vegetation over various lithological formations on Lesvos (Greece). Catena 40:51–68

Lasanta T, Garcia-Ruiz JM, Perez-Rontome C, Sancho-Marcen C (2000) Runoff and sediment yield in a semi-arid environment: the effect of land management after farmland abandonment. Catena 38:265–278

Lesschen JP, Kok K, Verburg PH, Cammeraat LH (2007) Identification of vunerable areas for gully erosion under different scenarios of land abandonment in Southeast Spain. Catena 71(1):110–121

Lesschen JP, Cammeraat LH, Kooijman AM, Van Wesemael B (2008a) Development of spatial heterogeneity in vegetation and soil properties after land abandonment. J Arid Environ 72(11):2082–2092

Lesschen JP, Cammeraat LH, Nieman T (2008b) Erosion and terrace failure due to agricultural land abandonment in a semi-arid environment. Earth Surf Process Landf 33(10):1574–1584

Lesschen JP, Schoorl JM, Cammeraat LH (2009) Modelling runoff and erosion for a semi-arid catchment based on hydrological connectivity to integrate plot and hillslope scale influences. Geomorphology 109:174–183

Levine CM, Stromberg JC (2001) Effects of flooding on native and exotic plant seedlings: implications for restoring south-western riparian forests by manipulating water and sediment flows. J Arid Environ 49:111–131

Lexartza-Artza I, Wainwright J (2009) Hydrological connectivity: linking concepts with practical implications. Catena 79(2):146–152

Li Y, Zhu X, Tian J (1991) Effectiveness of plant roots to increase the anti-scourability of soil on the loess plateau. Chin Sci Bull 36:2077–2082

Lioubimtseva E, Cole R, Adams JM, Kapustin G (2005) The impacts of climate and land-cover changes in arid lands of Central Asia. J Arid Environ 62:285–308

Lite SJ, Bagstad KJ, Stromberg JC (2005) Riparian plant species richness along lateral and longitudinal gradients of water stress and flood disturbance, San Pedro River, Arizona, USA. J Arid Environ 63:785–813

Ludwig JA, Tongway DJ (2000) Viewing rangelands as landscape systems. In: Archer S, Arnold O (eds) Rangeland desertification. Kluwer Academic Publishers, Dordrecht, pp 39–76

Ludwig JA, Eager RW, Gary NB, Chewings VH, Liedloff AC (2002) A leakiness index for assessing landscape function using remote sensing. Landsc Ecol 17:157–171

Maestre FT, Cortina J (2004) Are Pinus halepensis plantations useful as a restoration tool in semi-arid Mediterranean areas? For Ecol Manag 198:303–317

Malard F, Tockner K, Dole-Olivier M-J, Ward JV (2002) A landscape perspective of surface-subsurface hydrological exchanges in river corridors. Freshw Biol 47:621–640

Mamo M, Bubenzer GD (2001a) Detachment rate, soil erodibility and soil strength as influenced by living plant roots, part I: laboratory study. Am Soc Agric Eng 44:1167–1174

Mamo M, Bubenzer GD (2001b) Detachment rate, soil erodibility and soil strength as influenced by living plant roots, Part II: field study. Am Soc Agric Eng 44:1175–1181

Mant J (2002) Vegetation in the ephemeral channels of southeast Spain: its impact on and response to morphological change. PhD, University of Portsmouth, Portsmouth, United Kingdom

Marchamalo M, Hooke JM, Sandercock PJ (2016) Flow and sediment connecticity in semi-arid landscapes in SE Spain: patterns and controls. Land Degrad Dev 27:1032–1044

Martínez-Casasnovas JA, Sánchez-Bosch I (2000) Impact assessment of changes in land use/conservation practices on soil erosion in the Penedès-Anoia vineyard region (NE Spain). Soil Tillage Res 57:101–106

Martínez-Fernández J, Esteve MA (2005a) A critical view of the desertification debate in Southeastern Spain. Land Degrad Dev 16:529–539

Martínez-Fernández J, Esteve MA (2005b) A critical view of the desertification debate in Southeastern Spain. Land Degrad Dev 16:529–539

Martinez-Fernandez J, Lopez-Bermudez F, Romero-Díaz MA (1995) Land use and soil-vegetation relationships in a Mediterranean ecosystem, El Ardal, Murcia, Spain. Catena 25:53–167

Mattia C, Bischetti GB, Gentile F (2005) Biotechnical characteristics of root systems of typical Mediterranean species. Plant Soil 278:23–32

McAlpine KG, Drake DR (2002) The effects of small-scale environmental heterogeneity on seed germination in experimental treefall gaps in New Zealand. Plant Ecol 165:207–215

McBride JR, Strahan J (1984) Establishment and survival of woody riparian species on gravel bars of intermittent streams. Am Midl Nat 112:235–245

Meerkerk AL, van Wesemael B, Cammeraat LH (2008) Water availability in almond orchards on marl soils in south east Spain: the role of evaporation and runoff. J Arid Environ 72:2168–2178

Meerkerk AL, van Wesemael B, Bellin N (2009) Application of connectivity theory to model the impact of terrace failure on runoff in semi-arid catchments. Hydrol Process 23(19):2792–2803

Metochis C, Orphanos PI (1997) Yield of barley under Mediterranean conditions of variable rainfall. Agric For Meteorol 85:251–258

Micke WC (1996) Almond production manual. University of California, Division of Agriculture and Natural Resources, Oakland

Millenium Ecosystem Assessment (2005) Ecosystems and human well-being: desertification synthesis. In World Resources Institute (ed), Washington DC, p 28

Mitchell TD, Hulme M, New M (2002) Climate data for political areas. Area 34:109–112

Morgan RPC (ed) (2005) Soil erosion and conservation, 3rd edn. Blackwell Science Ltd, Oxford (UK), p 299

Moro MJ, Domingo F (2000) Litter decomposition in four woody species in a Mediterranean climate: weight loss, N and P dynamics. Ann Bot 86:1065–1071

Motha JA, Wallbrink PJ, Hairsine PB, Grayson RB (2004) Unsealed roads as suspended sediment sources in an agricultural catchment in south-eastern Australia. J Hydrol 286:1–18

Navarro-Cano JA (2004) Plant species and vegetation in channels and gullies in Cárcavo Basin. Internal Report for RECONDES Project

Navarro-Cano JA, Barberá GG, Ruiz-Navarro A, Castillo VM (2009) Pine plantation bands limit seedling recruitment of a perennial grass under semiarid conditions. J Arid Environ 73(1):120–126

Navarro-Cano JA, Barberá GG, Castillo VM (2010) Pine litter from afforestations hinders the establishment of endemic plants in semiarid scrubby habitats of Natura 2000 network. Restor Ecol 18(2):165–169

Norris JE (2005) Root reinforcement by hawthorn and oak roots on a highway cut-slope in Southern England. Plant Soil 278:43–53

O'Day CM, Phillips JV (2000) Computed roughness coefficients for Skunk Creek above Interstate 17, Maricopa county, Arizona Water-Resources Investigations Report 99—4248, U.S. Department of the Interior, U.S. Geological Survey, p 25

Okin GS, Parsons AJ, Wainwright J, Herrick JE, Bestelmeyer BT, Peters DC, Fredrickson EL (2009) Do changes in connectivity explain desertification? Bioscience 59(3):237–244

Operstein V, Frydman S (2000) The influence of vegetation on soil strength. Ground Improv 4:81–89

Osterkamp WR, Hupp CR (1984) Geomorphic and vegetative characteristics along three northern Virginia streams. Geol Soc Am Bull 95(9):1093–1101

Pain ADM (1985) Ergodic' reasoninhg in geomorphology: time for a review of the term? Prog Phys Geogr 9:1–15

Pastor M (2004) Sistemas de manejo del suelo. In: Barranco D, Fernández-Escobar R, Rallo L (eds) El cultivo del olivo. Ediciones Mundi-Prensa, Madrid, pp 229–285

Phillips JV, Hjalmarson HW (1994) Floodflow effects on riparian vegetation in Arizona. In: Cotroneo George V, Rumer Ralph R (eds) Hydraulic engineering '94, vol 1994. American Society of Civil Engineers. Hydraulics Division, New York, pp 707–711

Phillips JV, Ingersoll TL (1998) Verification of roughness coefficients for selected natural and constructed stream channels in Arizona. U. S. Geological Survey, Reston

Phillips JV, McDoniel D, Capesius JP, Asquith W (1998) Method to estimate effects of flow-induced vegetation changes on channel conveyances of streams in central Arizona Water-Resources Investigations Report 98–4040, U.S. Department of the Interior, U.S. Geological Survey, p 43

Pinto-Correia T, Mascarenhas J (1999) Contribution to the extensification/intensification debate: new trends in the Portuguese montado. Landsc Urban Plan 46:125–131

Poesen JWA, Hooke JM (1997) Erosion, flooding and channel management in Mediterranean environments of southern Europe. Prog Phys Geogr 21(2):157–199

Poesen J, Hooke J (1999) Erosion, flooding and channel management in desertification prone areas of the European Mediterranean. Paper presented at the Mediterranean desertification: research results and policy implications, Crete, Greece

Poesen J, Valentin C (2003) Gully erosion and global change. Catena 50(2–4):87–562

Poesen J, De Luna E, Franca A, Nachtegaele J, Govers G (1999) Concentrated flow erosion rates as affected by rock fragments cover and initial soil moisture content. Catena 36:315–329

Poesen J, Nachtergaele J, Verstraeten G, Valentin C (2003) Gully erosion and environmental change: importance and research needs. Catena 50:91–133

Prescott CE, Blevins LL, Stanley CL (2004) Litter decomposition in B.C. forests: controlling factors and influences of forestry activities. Br Columbia J Ecosyst Manage 5:44–57

Puigdefabregas J, Sole A, Gutierrez L, del Barrio G, Boer M (1999) Scales and processes of water and sediment redistribution in drylands: results from the Rambla Honda field site in Southeast Spain. Earth Sci Rev 48(1–2):39–70

Querejeta JI (1998) Efectos del tratamiento combinado de suelo y planta sobre una repoblación de Pinus halepensis Mill. en ambiente semiárido. PhD, Universidad de Murcia, Murcia, Spain

Querejeta JI, Roldán A, Albaladejo J, Castillo V (2001) Soil water availability improved by site preparation in a Pinus halepensis afforestation under semiarid climate. For Ecol Manag 149(1–3):115–128

Quiñonero-Rubio JM, Boix-Fayos C, de Vente J (2013) Desarrollo y aplicación de un índice multifactorial de conectividad de sedimentos a escala de cuenca. Cuadernos de Investigación Geográfica 39(2):203–223

Quinton JN, Morgan RPC, Archer NA, Hall GM, Green A (2002) Bioengineering principles and desertification mitigation. In: Geeson NA, Brandt CJ, Thornes JB (eds) Mediterranean desertification: a mosaic of processes and responses. Wiley, Chichester

Raes D, Geerts S, Kipkorir E, Wellens J, Sahli A (2006) Simulation of yield decline as a result of water stress with a robust soil water balance model. Agric Water Manag 81:335–357

Rallo L, Cuevas J (2004) Fructificatión y producción. In: Barranco D, Fernández-Escobar R, Rallo L (eds) El cultivo del olivo. Ediciones Mundi-Prensa, Madrid, pp 125–158

Ramos MC, Martínez-Casasnovas JA (2006) Impact of land levelling on soil moisture and runoff variability in vineyards under different rainfall distribution in a Mediterranean climate and its influence on crop productivity. J Hydrol 321:131–146

Reaney SM, Bracken LJ, Kirkby MJ (2007) Use of the connectivity of runoff model (CRUM) to investigate the influence of storm characteristics on runoff generation and connectivity in semi-arid areas. Hydrol Process 21:894–906

Rey F (2003) Influence of vegetation distribution on sediment yield in forested marly gullies. Catena 50:549–562

Rey F (2004) Effectiveness of vegetation barriers for marly sediment trapping. Earth Surf Process Landf 29:1161–1169

Rey F (2006) Vegetation dynamics on sediment deposits upstream of bioengineering works in mountainous marly gullies in a Mediterranean Climate (Southern Alps, France). Plant Soil 278:149–158

Rice EL (1979) Allelopathy: an update. Bot Rev 45:15–109

Rillig MC, Aguilar-Trigueros CA, Bergmann J, Verbruggen E, Veresoglou SD, Lehmann A (2015) Plant root and mycorrhizal fungal traits for understanding soil aggregation. New Phytol 205:1385–1388

Rivas-Martínez S (1987) Memoria del mapa de series de vegetación de España. ICONA, Madrid

Rodríguez LF (2006) Can invasive species facilitate native species? Evidence of how, when, and why these impacts occur. Biol Invasions 8(4):927–939

Rossi M, Torri D, Diele F, Marangi C, Ragni S, Blonda P, Nagendra H, Marchesini I, Santi E (2013) Report on habitat state and ecosystem status assessment: FP7-SPA-2010-1-263435 BIO_SOS deliverable D6.5 http://www.biosos.eu/deliverables/D6-5.pdf

Rossi M, Torri D, Santi E (2015) Bias in topographic thresholds for gully heads. Nat Hazards 79:S51–S69

Ruiz-Navarro A, Barberá GG, Navarro-Cano JA, Castillo VM (2009) Soil dynamics in Pinus halepensis reforestation: effect of microenvironments and previous land use. Geoderma 153(3–4):353–361

Ruiz-Navarro A, Barberá GG, García-Haro J, Albaladejo J (2012) Effect of the spatial resolution on landscape control of soil fertility in a semiarid area. J Soils Sediments 12(4):471–485

S, DB, Quine TA, Poesen J (2014) Root strategies for rill and gully erosion control. In Morte A, Varma A (eds) Root engineering: basic and applied concepts, Vol. Soil Biology Vol 40. Springer-Verlag, Berlin Heidelberg, p 493

Safriel U, Adeel Z (2005) Dryland systems. In: Hassan R, Scholes R, Ash N (eds) Ecosystems and human well-being: current state and trends. Island Press, Washington DC, pp 625–656

Sánchez Gómez P, Guerra Montes J (2003) Nueva flora de Murcia – Plantas vasculares. Libero Editor, Murcia

Sandercock PJ, Hooke JM (2006) Strategies for reducing sediment connectivity and land degradation in desertified areas using vegetation: the RECONDES project. In: Rowan JS, Duck RW, Werritty A (eds) Sediment dynamics and hydromorphology of fluvial systems, IAHS Publn. 306. IAHS, Wallingford, pp 200–206

Sandercock PJ, Hooke JM (2010) Assessment of vegetation effects on hydraulics and of feedbacks on plant survival and zonation in ephemeral channels. Hydrol Process 24(6):695–713

Sandercock PJ, Hooke JM (2011) Vegetation effects on sediment connectivity and processes in an ephemeral channel in SE Spain. J Arid Environ 75(3):239–254

Sandercock PJ, Hooke JM, Mant JM (2007) Vegetation in dryland channels and its interaction with fluvial processes. Prog Phys Geogr 31(2):107–129

Smit AL, Bengough AG, Engels C, van Noordwijk M, Pellerin S, van de Geijn S (eds) (2000) Root methods: a handbook. Springer-Verlag, Berlin, p 587

Smith WA, Dodd JL, Skinner QD, Rodgers JD (1993) Dynamics of vegetation along and adjacent to an ephemeral channel. J Range Manag 46:56–64

Snogerup S (1993) A revision of Juncus subgen. Juncus (Juncaceae). Willdenowia 23(1–2):23–73

Tabacchi E, Planty-Tabacchi AM, Salinas MJ, Decamps H (1996) Landscape structure and diversity in riparian plant communities: a longitudinal comparative study. Regul Rivers Res Manag 12:367–390

Thornes JB, Brandt J (1993) Erosion–vegetation competition in a stochastic environment undergoing climatic change. In: Millington AC, Pye KJ (eds) Environmental change in the drylands: biogeographical and geomorphological responses. Chichester, Wiley, pp 306–320

Tongway D, Ludwig B (1996) Rehabilitation of semiarid landscapes in Australia. II restoring productive soil patches. Restor Ecol 4:388–397

Tongway DJ, Sparrow AD, Friedel MH (2003) Degradation and recovery processes in arid grazing lands of central Australia. Part 1: soil and land resources. J Arid Environ 55(2):301–326. doi:10.1016/s0140-1963(03)00025-9

Tormo J, Bochet E, García-Fayos P (2007) Roadfill revegetation in semiarid Mediterranean environments. Part II: topsoiling, species selection, and hydroseeding. Restor Ecol 15(1):97–102

Torri D, Poesen J (2014) A review of topographic threshold conditions for gully head development in different environments. Earth Sci Rev 130:73–85

Tosi M (2007) Root tensile strength relationships and their slope stability implications of three shrub species in Northern Apennines (Italy). Geomorphology 87(4):268–283

Tubeileh A, Bruggeman A, Turkelboom F (2004a) Growing olive tree species in marginal dry environments. ICARDA, Aleppo

Tubeileh A, Bruggeman A, Turkelboom F (2004b) Growing olive tree species in marginal dry environments. ICARDA, Aleppo

UNCCD (1994) Convention to combat desertification. United Nations

UNEP (1997) World atlas of desertification. Arnold, London

van Dijk PM, Auzet AV, Lemmel M (2005) Rapid assessment of field erosion and sediment transport pathways in cultivated catchments after heavy rainfall events. Earth Surf Process Landf 30(2):169–182

Vannoppen W, Vanmaercke M, De Baets S, Poesen J (2015) A review of the mechanical effects of plant roots on concentrated flow erosion rates. Earth Sci Rev 150:666–678

Verheijen F, Cammeraat LH (2007) The association between three dominant shrub species and water repellent soils along a range of soil moisture contents in semi-arid Spain. Hydrol Process 21:2310–2316

Vigiak O, Borselli L, Newham LTH, Mcinnes J, Roberts AM (2012) Comparison of conceptual landscape metrics to define hillslope-scale sediment delivery ratio. Geomorphology 138:74–88

Vijfhuizen S (2005) Inventory for interactions between vegetation and surface: Carcavo basin, SE Spain. Student report. University of Amsterdam, Amsterdam

Wainwright J, Turnbull L, Ibrahim TG, Lexartza-Artza I, Thornton SF, Brazier RE (2011) Linking environmental regimes, space and time: interpretations of structural and functional connectivity. Geomorphology 126(3–4):387–404

Western AW, Grayson RB, Blöschl G, Willgoose GR, McMahon TA (1999) Observed spatial organization of soil moisture and its relation to terrain indices. Water Resour Res 35:797–810

Williams A, Ternan JL, Elmes A, González del Tánago M, Blanco R (1995) A field study of the influence of land management and soil properties on runoff and soil loss in Central Spain. Environ Monit Assess 37:333–345

Wolman MG, Gerson R (1978) Relative scales of time and effectiveness of climate in watershed geomorphology. Earth Surf Process Landf 3:189–208

Xiong S, Nilsson C (1999) The effect of plant litter on vegetation: a meta-analysis. J Ecol 87:984–994

Yin YH, Wu SH, Zheng D, Yang QY (2005) Regional difference of aridity/humidity conditions change over China during the last thirty years. Chin Sci Bull 50:2226–2233

Zhang J, Tian G, Li Y, Lindstrom M (2002) Requirements for success of reforestation projects in a semiarid low-mountain region of the Jinsha River basin, southwestern China. Land Degrad Dev 13:395–401

Zimmerman RC (1969) Plant ecology of an arid basin, Tres Alamos-Redington area south-eastern Arizona. United States Government Printing Office, Washington

Zimmerman RC, Thom BG (1982) Physiographic plant geography. Prog Phys Geogr 6:45–59

Zollo AL, Rillo V, Bucchignani E, Montesarchio M, Mercogliano P (2015) Extreme temperature and precipitation events over Italy: assessment of high-resolution simulations with COSMO-CLM and future scenarios. Int J Climatol 36(2):987–1004

Printed in the United S...
by Bookma... ...

Printed in the United States
By Bookmasters